Springer Theses

Recognizing Outstanding Ph.D. Research

Aims and Scope

The series "Springer Theses" brings together a selection of the very best Ph.D. theses from around the world and across the physical sciences. Nominated and endorsed by two recognized specialists, each published volume has been selected for its scientific excellence and the high impact of its contents for the pertinent field of research. For greater accessibility to non-specialists, the published versions include an extended introduction, as well as a foreword by the student's supervisor explaining the special relevance of the work for the field. As a whole, the series will provide a valuable resource both for newcomers to the research fields described, and for other scientists seeking detailed background information on special questions. Finally, it provides an accredited documentation of the valuable contributions made by today's younger generation of scientists.

Theses are accepted into the series by invited nomination only and must fulfill all of the following criteria

- They must be written in good English.
- The topic should fall within the confines of Chemistry, Physics, Earth Sciences, Engineering and related interdisciplinary fields such as Materials, Nanoscience, Chemical Engineering, Complex Systems and Biophysics.
- The work reported in the thesis must represent a significant scientific advance.
- If the thesis includes previously published material, permission to reproduce this must be gained from the respective copyright holder.
- They must have been examined and passed during the 12 months prior to nomination.
- Each thesis should include a foreword by the supervisor outlining the significance of its content.
- The theses should have a clearly defined structure including an introduction accessible to scientists not expert in that particular field.

More information about this series at http://www.springer.com/series/8790

Bettina Sellmeier

Quantitative Parameterization and 3D-run-out Modelling of Rockfalls at Steep Limestone Cliffs in the Bavarian Alps

Doctoral Thesis accepted by
Technische Universität München, Germany

Springer

Author
Dr. Bettina Sellmeier
Technische Universität München
Munich
Germany

Supervisor
Prof. Kurosch Thuro
Technische Universität München
Munich
Germany

ISSN 2190-5053 ISSN 2190-5061 (electronic)
Springer Theses
ISBN 978-3-319-30586-8 ISBN 978-3-319-24510-2 (eBook)
DOI 10.1007/978-3-319-24510-2

Springer Cham Heidelberg New York Dordrecht London

Softcover re-print of the Hardcover 1st edition 2015

Printed on acid-free paper

Springer International Publishing AG Switzerland is part of Springer Science+Business Media
(www.springer.com)

Parts of this thesis have been published in the following articles:

2011

SELLMEIER, B., & THURO, K. (2011): Possibilities and limitations of 2d and 3d rockfall-simulations concerning the digital terrain model (DTM). - In: Interdisciplinary workshop on rock fall protection 2011. - Congresspark Igls, Innsbruck May 17th to 19th, 2011, 5 p. (USB-Stick).

SELLMEIER, B. & THURO, K. (2011): Possibilities and limitations of 2D and 3D rockfall-simulations concerning the Digital Terrain Model (DTM). – In MARSCHALLINGER, R. & ZOBL, F. (ed.): IAMG-Mathematical Geosciences at the Crossroads of Theory and Practice, September 5th to 9th, 2011, Salzburg. – 5 S. (Paper Nummer: doi: 10.5242, USB Stick).

SELLMEIER, B. & THURO, K. (2011): Possibilities and limitations of 2D and 3D rockfall modeling—finding an optimal way for integrating protection systems.—In: Slope stability 2011. - International Symposium on Rock Slope Stability in Open Pit Mining and Civil Engineering, Vancouver, Canada September 18th to 21st, 2011, 7 p. (USB-Stick).

2012

SELLMEIER, B., KRAUTBLATTER, M. & THURO, K. (2012): Possibilities and limitations of 2D and 3D Rockfall Simulations—which kinds of uncertainties do we have to consider? - Conference of the European Geoscience Union (EGU), April 22nd to 27th, 2012, (Vienna).

2013

SELLMEIER, B., KRAUTBLATTER, M. & THURO, K. (2013): Enhanced hazard assessment of a steep limestone rock slope above the federal road B 305. - In: Thuro, K. (eds.): 19. Tagung für Ingenieurgeologie mit Forum für junge Ingenieurgeologen. – 660 S., Technische Universität München, March 13th to 15th, 2013, München.

SELLMEIER, B., ZUMBRUNNEN, TH., KRAUTBLATTER, M. & THURO, K. (2013): Mid magnitude rockfalls: too big for protection measures - too small for acceleration anticipation? - Österreichische Gesellschaft für Geomechanik, Workshop Versagensprognose in der Geotechnik, October 9th, 2013.

SELLMEIER, B., KRAUTBLATTER, M. & THURO, K. (2013): Enhanced hazard assessment of a steep limestone rock slope above the federal road B 305. - Conference of the European Geoscience Union (EGU), April 7th to 12th, 2013, (Vienna).

2014

Sellmeier, B., Krautblatter, M. & Thuro, K. (2014): Quantitative parameterization of mid-magnitude rockfalls at a sensitive stage of failure in the Bavarian Alps (Berchtesgaden). - Conference of the European Geoscience Union (EGU), April 27th to May 2nd (Vienna).

Sellmeier, B., Krautblatter, M. & Thuro, K. (2014): Failure and mobilization analysis of mid-magnitude rockfalls on a steep limestone slope in the Bavarian Alps. - International Association for Engineering Geology (IAEG), proceedings, IAEG congress September 15th to 19th (Torino).

Sellmeier, B, Zumbrunnen, Th., Thuro, K. & Krautblatter, M. (2014): Steinschlagereignisse mittlerer Magnitude: Wie können wir eine Gefährdungs- und Prozessbeurteilung gewährleisten? - Wildbach- und Lawinenverbau, Heft 173, Verein der Diplomingenieure der Wildbach- und Lawinenverbauung Österreichs, 228–245.

Supervisor's Foreword

The Ph.D. thesis of Mrs. Bettina Sellmeier deals with the determination of input parameters for rockfall modelling in 3D and their application at a study site in the region of Berchtesgaden (Bavarian Alps, Germany). The project was set up in cooperation with the Bavarian State Agency for Environment and the State Agency of Road Construction in Traunstein, Germany. The objective was not only to consider regular low-magnitude, high-frequency rockfalls, but also mid-magnitude events, which are up to now insufficiently considered in 3D rockfall modelling.

2D rockfall run-out modelling is set to be a state-of-the-art tool for several years, but only for rock slope heights up to 40 m. The parameterization of rockfall scenarios is still demanding for the different codes and potential uncertainties increase with an increasing area of survey. At the current study site, the vertical height from the release area to federal road at the run-out area is about 600 m. For the investigation of the process areas (source, transit and run-out area), Bettina Sellmeier was able to set up innovative approaches, which incorporate not only low magnitudes (<10 m^3) but also mid-magnitudes (boulder falls 10–100 $m^3$and block falls up to 10,000 m^3). The latter exceed the load capacities of common mitigation measures in alpine regions. In this project, a geomechanical approach to characterize mid-magnitude events by means of a case study including a 200-m^3 block at the so-called Wachterl-Horn at the source area of the project site has been developed. For this block, a potential failure scenario was analysed by linking a deterministic failure analysis with a numerical process-based run-out model.

For the probabilistic models, 2D codes, in contrast to 3D codes, are already the state-of-the-art tools. The run-out analysis for the entire project site was performed using the modelling code Rockyfor3D (Dorren/ecorisQ). For modelling a potential run-out scenario of the 200-m^3 block, Bettina Sellmeier used the beta version of the code RAMMS::Rockfall, developed by the Swiss Institute for Snow and Avalanche Research (SLF). RAMMS::Rockfall provides the possibility of considering the block shape and thus relating different block shapes to the run-out distance. By means of this approach, the information about the discontinuities at the source area and the block shapes at the talus slope can be transferred into the rockfall model.

Due to a scientific cooperation between the Chair for Engineering Geology (Technische Universität München) and the SLF (Dr. Perry Bartelt), Bettina Sellmeier was able to contribute to the improvement and processing of the beta version of RAMMS::Rockfall.

The current Ph.D. thesis represents the first major project, in which RAMMS:: Rockfall was systematically applied and tested. At the same time, the Ph.D. thesis of James Glover, which deals with the processing of the rockfall code, was supervised by Prof. Dave Petely (University of Durham) in cooperation with the SLF (Dr. Perry Bartelt).

Although Bettina Sellmeier used 2D codes in the very beginning of her Ph.D. research, these preliminary studies were not taken into account for the final thesis. Thus, the current application of 3D rockfall modelling could be considered as future state of the art.

In this context, the aims of the current research project were given as follows:

1. The analysis of failure mechanics at rockfall source areas under consideration of crack generation and propagation in carbonate rocks. The study object was a 200-m^3 limestone block subjected to planar failure at the source area of the project site ("Wachterl-Horn"). The field work included the in-situ investigation of the failure surface due to the unique accessibility of a "failure cave" underneath the block.
2. The outline of a reproducible recommendation, summing up quantitative methods for determining the crucial discontinuities at the source area (influencing the maximum block sizes and magnitudes) as well as the geometric block inventory at the talus slope. This approach leads, especially for the current study, to a quantitative parameterization of the input parameters; not estimating the parameters like it is usually often the state of practice.
3. The performance of run-out modelling, analysing the influence of the slope roughness and damping parameters as well as the effect of windblow-areas on the run-out distance. The effect of potential block-fragmentation along the rockfall trajectories was approached for the case study of the critical 200 m^3 block.

The aspect of fragmentation can still be seen as an unsolved task in the field of applied rockfall modelling. The block sizes are usually determined by the intersection of discontinuity sets at the source area. The released blocks will be disintegrated and destroyed during the ground contacts along the falling path (trajectory); this process can be named as fragmentation. Bettina Sellmeier was able to give enhanced input to this topic due to a statistical evaluation of the block inventory at the talus slope, where the block dimensions were quantitatively recorded and taken into account for the run-out analysis. Up to now, the detailed quantitative recording was not taken into account for the rockfall modelling of the Bavarian State Agency for Environment or their consultants (GEOTEST, Switzerland). The latter is one of the main contractors in professional 3D rockfall-modelling projects for public clients like the Bavarian State Agency for Environment in Germany.

Up to the current state of the art, it is not possible to take the mentioned fragmentation scenarios into account for 3D rockfall codes. Nevertheless, using the code RAMMS::Rockfall, Bettina Sellmeier was able to set up various parameter studies analysing the effect of fragmentation processes on the run-out distance in terms of the current case study. The parameter studies were based on certain assumptions concluded from accurate field investigation.

Munich, Germany
June 2015

Prof. Kurosch Thuro

Acknowledgments

- Univ. Prof. Dr. rer. nat Kurosch Thuro: Thank you so much for the trust in me to work on the chosen Ph.D. topic, for the boundless support, the great discussions and most for being my doctoral father!
- Univ. Prof. Dr. rer. nat Michael Krautblatter: Thank you for supporting my work from the first moment you arrived in Munich, for the inspiring meetings and the great apple pies!
- Univ. Prof. Ph.D. Daniel Scott Kieffer: Thank you for the input to my work the helpful discussions and the time you invested for my Ph.D. project.
- Univ. Prof. Dr. rer.-nat. Christian Grosse: For the great support during the field trip to the block at the Wachterl-Horn and the perfect cooperation.
- Dr. Andreas v. Poschinger: For the interest in my project and the input from the Bavarian Agency for Environment
- Dr. Bernhard Krummenacher (GEOTEST): Thank you for the support in the starting stage of the Ph.D., for the meetings at the study site and the helpful discussions.
- Thomas Zumbrunnen (Stba Traunstein): For the idea to work on this stunning project site at the Weißwand and the tip of taking the critical block at the Wachterl-Horn into account!
- Prof. Dr. Perry Bartelt: Thank you very much for the interest in my research and for providing the RAMMS::Rockfall beta version.
- Dr. James Glover: Thank you so much for the support not only in terms of the model set-up but also in terms of discussing my results.
- Friedrich Ettl: You were a great help performing the test samples for the UCS and the direct shear tests! Thank you Fritz.
- Dr. Bernhard Lempe: Many thanks to my loyal office associate Bernie! You supported me not only with feedback concerning my figures but also with our chocolate "Flat-rate"!
- Dr. Katja Lokau: Thank you Katja for your support in terms of teaching during the final stage of my Ph.D.!
- Dr. Heiko Käsling: Thank you for your support during laboratory work and not at last for the funny times we had sharing an office.

- Silvia Beer, Judith Festl, Lisa Wilfing, Carola Wieser, Peter Ellecosta, Si-Yen Luu-Chucholowski and Philipp Mamot: Thank you guys for the great times and the encouragements during our lunch times and coffee breaks!
- Dr. Michael Rieder: Thank you for the encouragements and our chats about mountain climbing!
- Sonja Störzbach: Thanks to you Sonja for propping me up when I was down during stressful stages of the Ph.D. and in this context for the great times during climbing indoor and outdoor!
- Florian Menschik: Thank you so much Florian for your boundless support through all the ups and downs of the last years. You lightened up the demanding periods and made the lucky ones even brighter!
- My family: Last but not least I would like to thank my family from the bottom of my heart! You supported me in every circumstance throughout the whole period of the Ph.D. and beyond that!

Contents

Chapter 1
Introduction

Rockfall events of the last years like the one in Tramin at January 21st 2014 or the events at the Gotthard highway emphasize the significance of infrastructure rockfall-hazard-analysis. In the general context of natural hazards, risk is defined as the product of hazard and vulnerability:

$$R = H * V$$

where R is set to be the risk, H stands for hazard and V for vulnerability.

This formula takes the "hazard" in the context of natural hazards as well as the consequences (for example casualties and injuries) into account. The focus of the current thesis is set on the part of the "hazard" in this term, specifically the rockfall hazard along potentially affected road sections.

The approaches of rockfall hazard assessment are commonly based on regional modelling in 3D. Required input parameters for rockfall run-out modelling generally cover the slope parameters on the one hand and the parameters of the descending rock on the other hand. Especially the determination of the potential rock volumes can be demanding due to the hard accessibility of the rockfall source areas. The assessed rock volumes significantly influence the modelling results (the kinetic energies and the jump heights), which are approached as basic parameters for mitigation measure design. The high frequency low-magnitude rockfalls (<10 m^3, in total rock volume) constitute common events along infrastructure and therefor are well assessed in terms of run-out modelling. But there are potential events (also at the selected study site) exceeding the load capacities of mitigation measures due to the increased rock volume; consequently raising the question how to deal with such volume classes in terms of rockfall hazard assessment.

Up to now the way of natural hazard assessment in Bavaria is based on so called hazard indication maps (Gefahrenhinweiskarten) provided by the Bavarian State Office for the Environment (Bayerisches Landesamt für Umwelt). The indication maps cover the potential hazard of shallow—and deep landslides, collapse sinkholes and rockfalls. The datasets of the latter are based on rock volume classes, which are determined referring to the geological setting of different regions in the Bavarian Alps. Especially the road sections at the region Traunstein-Berchtesgaden in the south east of Bavaria (Germany) are regularly affected by rockfalls of varying

B. Sellmeier, *Quantitative Parameterization and 3D-run-out Modelling of Rockfalls at Steep Limestone Cliffs in the Bavarian Alps*, Springer Theses,
DOI 10.1007/978-3-319-24510-2_1

magnitudes. The federal road B 305, connecting the villages Unterjettenberg and Ramsau, is edged by steep valley flanks and rock cliffs providing potential rockfall material. The high traffic frequency of 2757 vehicles per day emphasizes the necessity of analysing the rockfall hazard along this road section (URL-01). In this context cooperation between the State Agency for Construction (Staatliches Bauamt Traunstein) and the Chair for Engineering Geology (Technische Universität München) was started to perform a detailed hazard analysis along the mentioned road section and its alpine vicinity.

The current Ph.D. project was embedded in the topic of rockfall hazard assessment including accurate field investigation and 3D run-out modelling, covering the following three main sections:

1. The detachment analysis of mid-magnitude rockfall by means of a case study: This part covers the detailed case study of an approximately 200 m^3 block located above a hiking track at the rockfall source area above the federal road B 305. By means of field investigation a mechanical interpretation of a potential detachment process as well as a stability analysis based on recorded shear parameters was performed. The unique occasion of an accessible failure surface underneath the block afforded the opportunity to record the joint roughness coefficient (JRC) and the joint compressive strength index (JCS) in situ at the failure surface. The recording of shear parameters was linked to a limit equilibrium analysis of the block subjected to planar failure.
2. The quantitative magnitude assessment for carbonate rockfalls, taking the talus material as well as the source area into account:
   This chapter deals with the quantitative assessment of rockfall volumes. On the one hand a quantitative recording of block dimensions at the talus slope was performed. On the other hand the discontinuities at the source area were analysed and in addition a kinematic analysis covering varying detachment processes was conducted. The potential block magnitudes were assessed in two ways: the minimum magnitudes were determined in sample areas of 20 by 20 m each, recording rocks at the talus slope and measuring the three block axes as well as the mean obstacle height (MOH) (Dorren 2012). The maximum magnitudes were determined by mapping the source area and performing scanline analysis. The results of the block magnitudes were used for the subsequent rockfall modelling.
3. The 3D rockfall modelling based on field investigation, using two different modelling approaches:
   The rockfall hazard assessment at the entire project site was analysed by means of a sensitivity analysis taking the rockfall volume as well as the slope roughness into account. Therefor a code based on an approach of the coefficients of restitution was applied (Rockyfor3D). For the critical 200 m^3 block described above the rockfall run-out for varying stages of fragmentation was assessed, using a modelling approach based on non-smooth contact dynamics (RAMMS:: Rockfall). Therefor an approach of modelling increased rockfall volumes (>10 m^3) is provided.

Summing up, the current research work aims to demonstrate how detailed field investigation could enhance the outcomes of rockfall run-out modelling. Besides varying stages of rockfall magnitudes were taken into account.

References

Dorren LKA (2012) Rocky for 3D (v5.0) revealed—Transparent description of the complete 3D rockfall model. ecorisQ paper (www.ecorisq.org) 31 p

URL-01. https://www.baysis.bayern.de/web/content/verkehrsdaten/SVZ/kennwerteundkarten.aspx?gebiet=bauamt&regbez=000&landkreis=000&bauamt=0916&tag=000. Access at 25 Feb 2015

Chapter 2
State of the Art

2.1 Characterisation of Rockfalls

According to the detachment process landslides can be divided in 5 the process-groups: slide, flow, drift, topple, and fall. Each of these process types can be combined with the type of material: rock, debris, soil or mud. This PhD thesis focuses the process of rockfalls, which implies the detachment of a rock at a steep slope surface (Cruden and Varnes 1996). The detached rock material descends the slope in the possible modes of motion namely falling, bouncing, rolling or sliding.

During the free fall two different modes of falling can occur: either the block rotates around its centre of mass or a translation of the centre of the block takes place (Azzoni et al. 1995: 712). If the slopes inclination angle falls below 76 degrees, falling transforms to bouncing. During the bouncing process the tangential and normal coefficients of restitution and the angle between the block-trajectory and the slope would affect the rebound behaviour most (Hungr and Evans 1988 in Dorren 2003). Below an inclination angle of 45° the block would continue rolling (Ritchie 1963 in Dorren 2003: 72). The velocity of the process "fall" can be classified as very rapid to extremely rapid, meaning 0.3 m/min to >3 m/s (Cruden and Varnes 1996).

Due to rock volume we can further classify the process of rockfall to debris fall (<10 m^3), boulder fall (10–100 m^3), block fall (>100 m^3), cliff fall (10^4–10^6 m^3) and Bergsturz (>10^6 m^3) increasing the volume (Whalley 1984: 218).

B. Sellmeier, *Quantitative Parameterization and 3D-run-out Modelling of Rockfalls at Steep Limestone Cliffs in the Bavarian Alps*, Springer Theses,
DOI 10.1007/978-3-319-24510-2_2

2.2 Towards Characterizing Rock Mass: Influencing Parameters and Classifications

The formation of cracks and later rock joints depends on the material properties of the rock, especially on the materials strength and on the induced stresses as well as their direction. In case of failure, the induced stresses exceed the materials strength.

Discontinuities represent interstices in the rock mass along which no tensile stress can be transferred (Priest 1993: 5). Different kinds of discontinuities can be divided referring to their origin or geological history (Priest 1993: 10 ff.). Faults develop through induced tectonically shear stress which exceeds the shear strength of the rock material along a particular plane (Kersten 1990 in Priest 1993). Faults in general are often grouped in sets, so called fault zones. A characteristic feature would be the core of a fault zone consisting of powdered rock (fault gouge or fault breccia), which is surrounded by the disintegrated rock mass. Joints represent fractures in a rock mass along which little or no displacement has occurred (Price 1966 in Priest 1993: 12). The joint frequency and occurrence is mostly controlled by lithology and bed thickness. This type of discontinuity can be divided in two groups: systematic joints, which mostly run in planar and at least sub-parallel joint sets and non-systematic joints, which cannot be grouped together in a sense of joint sets. Bedding planes are characterized by physical or chemical changes in the material during the deposition, for example grain size or dolomite content. These changes can, but do not have to lead to parallel or sub-parallel fractures drawing through the rock mass. The cleavage can be separated into two types: the fracture cleavage and the flow cleavage. The fracture cleavage is not connected to any kind of parallel aligned minerals and describes cemented parallel fractures in a rock mass. The flow cleavage is dependent on recrystallization of minerals and a parallel alignment of metamorphic minerals in a typical sense, like mica, leading to the classical type of foliation.

The discontinuities generally contribute to an increasing destabilisation of the rock mass meaning that the strength of a rock mass disintegrated through joints would always be less than the materials strength.

The Characterisation of Joints

The discontinuity pattern in a rock mass can be described in terms of the following criteria: the orientation, the persistence, the spacing and the surface conditions (Fig. 2.1). The mentioned objectives can be quantitatively recorded in the field due to scanline analysis using a fact sheet (Priest 1993: 33, Norrish and Wyllie 1996: 391). It is recommendable to additionally record the aperture width and the discontinuity filling, especially for geotechnical purposes (ISRM 1978: 351 ff.). The different parameters of discontinuities determine the degree of disintegration, the joint properties and subsequently the potential of failure. The objectives of crack development and the shear abilities of rock joints are described in Sect. 2.2.2 in more detail.

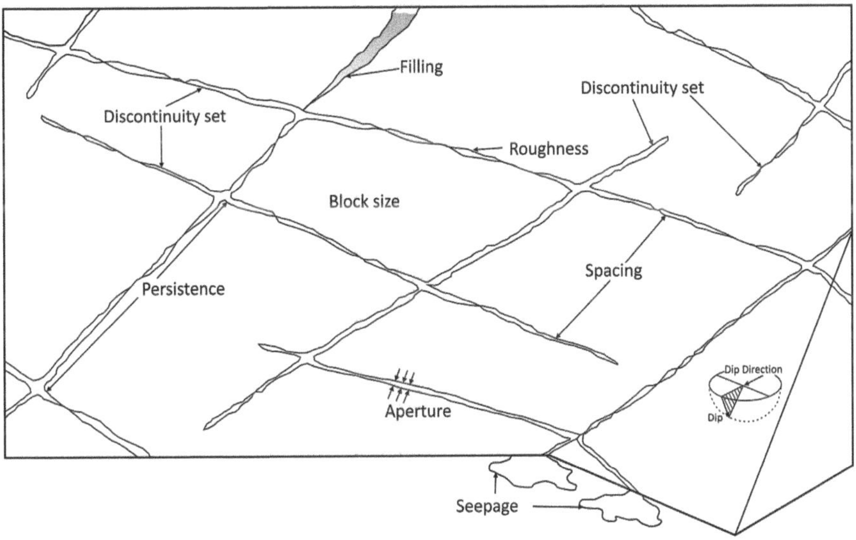

Fig. 2.1 Illustration of the joint parameters influencing the rock mass (in Hudson and Harrison 1997, p. 116)

The orientation of a discontinuity is usually measured as dip and dip-direction and can be visualized in a stereographic projection, for example in a Schmidt Net. By use of a Schmidt Net the spatial distribution of the discontinuities can be visualized in relation to each other. The discontinuity persistence describes the visible trace length of a discontinuity on a rock face above and below the location of recording, named as upper and lowers semi trace length (Hudson and Priest 1979; Priest 1993). The persistence can also be connected to a five step classification from very low persistence (<1 m) to very high persistence (>20 m) (ISRM 1978: 335). The challenge with persistence is the assessment of discontinuity progression in the internal of the rock mass. The spacing represents the distance between the discontinuities of one set (Fig. 2.2). The spacing can be determined along a scanline and afterwards be transformed to a true spacing, which represents the spacing measured perpendicular to the joint surfaces. The spacing and persistence of different joint sets have a main effect on the cubature of critical blocks.

The surface conditions include information about the fracture roughness, which could be determined by the joint roughness coefficient (JRC, 1 = very smooth to 20 = extremely rough; Fig. 2.3) (Barton 1973: 314 ff., Barton and Choubey 1977: 19) and the curvature or waviness of a discontinuity characterised in 5 classes, 1 = planar and 5 = very curved (Priest 1993: 44). The qualitative description of discontinuity surface roughness is characterized according to 3 main types (stepped, undulating or planar), each divided into the three subtypes: rough, smooth and slickensided (ISRM 1978: 343). The visual comparison of the given roughness profiles with the rock surfaces should usually be carried out in dip direction. For the

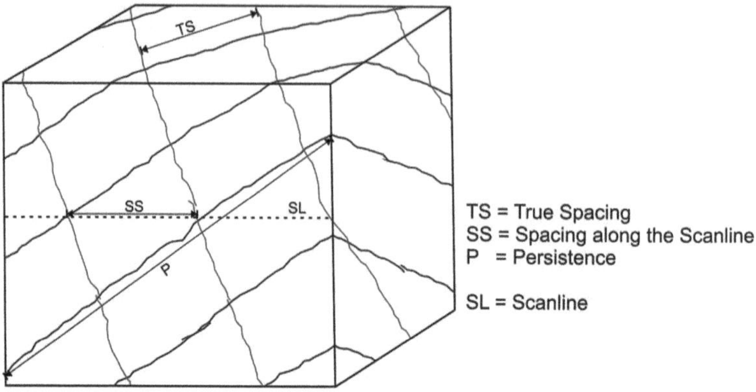

Fig. 2.2 The terms of persistence and spacing visualized in a block diagram

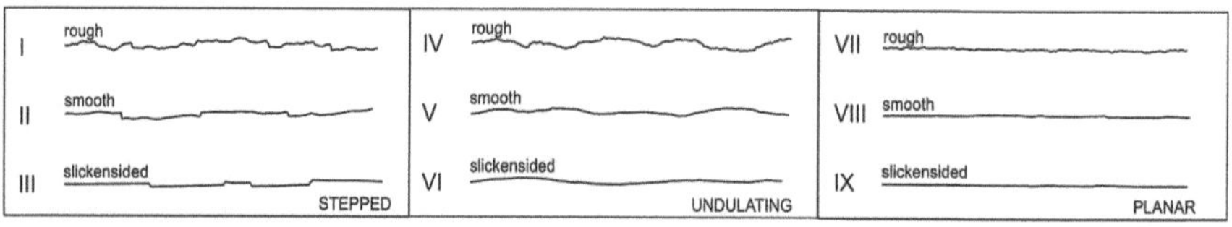

Fig. 2.3 Quantitative roughness classification categorized according to three main types: stepped, undulating and planar; each divided in three subtypes: rough, smooth and slickensided (after ISRM 1978: 343)

use of the subclasses termed “slickensided” a certain evidence for shear displacement has to be provided.

The aperture width is measured perpendicular to the rock walls of an opened joint. The aperture can be classified referring to three categories: closed features, gaped features and open features, which are subdivided in nine categories, from very tight (<0.1 mm) to cavernous (>1 m) (ISRM 1978: 352) (Table 2.1).

Table 2.1 Description of joint apertures referring to ISRM (1978: 352)

Aperture	Description	
<0.1 mm	Very tight	“Closed” features
0.1–0.25 mm	Tight	
0.25–0.5 mm	Partly open	
0.5–2.5 mm	Open	“Gapped” features
2.5–10 mm	Moderately wide	
>10 mm	Wide	
1–10 cm	Very wide	“Open” features
10–100 cm	Extremely wide	
>1 m	Cavernous	

It should be noted that a difference between the terms aperture and width exists due to the filling of joints. A joint filled with air or water shows an aperture whereas a joint filled with clay or other sedimentary material shows a width of opening (Fig. 2.1).

The infilling of a discontinuity represents the material separating two joint walls from each other. Factors like the filling material, the materials particle size, the water content, the joint wall roughness, the fracturing of the surrounding rock mass or the degree of weathering play an important role for the geotechnical behaviour of the filled discontinuities (ISRM 1978: 355 ff.).

Rock Mass Classification Systems

For classifying a rock mass originally most input was given in the field of tunneling (Terzaghi 1946 in Hoek 2000), where the rock mass was classified according to the rock properties and the degree of transection. The rock mass rating system (RMR) suggests the characterization of the rock mass referring to a score system, in which the geotechnical parameters are weighted according to their degree of influence on the rock mass destabilization (Bieniawski 1976 in Hoek 2000: 47). The Rock Quality Designation (RQD) gave input to the topic of rock mass classification using drilling cores and was actually linked to a rock mass classification system (Deere et al. 1969). The critical parameters of rock faces endangering certain highway sections can be assessed using the rockfall hazard rating system which was set up as a score system to characterize the effect on a certain highway section (Pierson 1991). The Slope Stability Probability Classification (SSPC) provides a three steps approach for characterizing a certain rock mass (Hack et al. 2002). The system includes the characterization of the rock mass (exposure rock mass, ERM) the characterization of the bedrock material (reference rock mass, RRM) and the stage of the slope after construction taking future weathering into account (slope rock mass, SRM). Nevertheless the slope and joint geometry affect the volume and shape of rockfall material. Since the joint persistence is most difficult to determine in situ, the effect of joint geometry and persistence on rock slope stability can be assessed via probabilistic approaches like SLOPESIM (Einstein et al. 1983).

2.2.1 The Mechanics of Failure Preparation and Causes

The Total Friction Concept

The geotechnical hazard analysis at rockfall source areas along the B 305 is performed using the total friction concept (Barton and Choubey 1977). The total friction concept takes the roughness of rock joints into account, when considering the Mohr-Coulomb failure criterion (Patton 1966 in Barton 1973, Goldstein et al. 1966 in Barton 1973).

Thus, the shear stress at failure τ is approached by

$$\tau = \sigma * \tan(\varphi_b + i) \tag{2.1}$$

where σ is the normal stress, φ_b is the material-immanent basic friction angle and i is the dilation angle that expresses effective roughness. The friction term is composed of two parameters, the basic friction angle φ_b and the dilation angle i of the joint surface. The basic friction angle is defined as the friction angle determined by shearing two planar rough "sawn or sand blasted" rock surfaces (Barton 1973, 1976; Barton and Choubey 1977).

The dilation angle i represents the asperity angle from the horizontal. As i is difficult to parameterise, Barton (1973) modified this equation for the shear stress at failure,

$$\tau = \sigma * \tan\left[JRC * \log\frac{JCS}{\sigma} + \varphi_b\right] \tag{2.2}$$

where JCS is the joint wall compressive strength and JRC is the joint roughness coefficient.

For rough and undulating surfaces the envelope of the peak shear strength shows a curved character (Barton 1973; Barton and Choubey 1977).

The roughness of rock joints can be categorized in first and second order asperities, depending on the considered scale (Selby 1982). The joint roughness coefficient can be assumed as a parameter for second order asperities since undulation is not taken into account. The **JRC** can be quantified using two different approaches. First the roughness of natural rock joints can visually be compared to 10 standardized profiles with JRC values from 0 (very smooth) to 20 (very rough) (Barton and Choubey 1977; Fig. 2.4).

More reproducible, the direct quantitative (in situ) method is assessing the JRC by using a profile gauge (Barton and Choubey 1977). The JRC-value is computed via the first derivation of roughness amplitudes (the gradient Z_2), which are determined in defined intervals of length (Tse and Cruden 1979; McCarroll 1997). Using linear regression analysis the JRC can be determined via Z_2 (Tse and Cruden 1979). The work of Yang et al. (2001) is based on the findings of Tse and Cruden (1979) and provides a better correlation between Z_2 and the JRC for the regression analysis. Theoretically, the JRC can be determined in a back-analysis of direct shear tests, assuming that shear strength, normal stress as well as JCS and basic friction angle are measured (Barton and Choubey 1977).

The JCS can be approached by both, uniaxial compressive strength (UCS) testing in the laboratory and Schmidt-Hammer in situ testing. UCS is definitely the more accurate test of material performance. However, Schmidt Hammer testing is better capable of defining spatial heterogeneity along rock joints with different degrees of weathering and thus material performance of rock bridges (Barton 1973: 328). For unweathered samples or rock joints the JCS is equivalent to the uniaxial compressive strength of the material (Barton 1973; Barton and

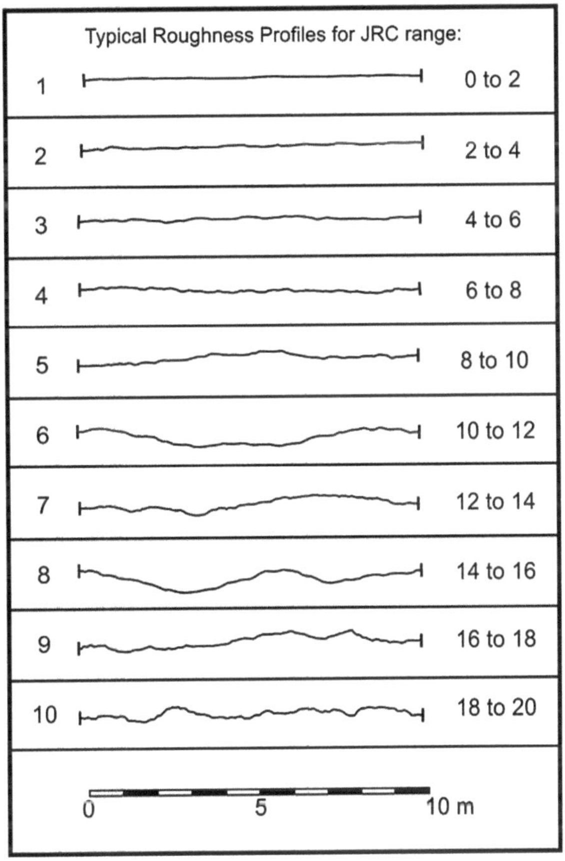

Fig. 2.4 Typical JRC profiles for comparing fracture roughness of rock faces with the graphical classification

Choubey 1977). For weathered material, the JCS is reduced in comparison to the UCS and shows a wider scatter (Barton 1973). In addition, surface roughness and moisture content contribute to a reduction of the JCS (Aydin and Basu 2005; Barton and Choubey 1977; Goudie 2006).

Causes

The causes are landslide promoting factors which do not instantaneously lead to failure. The causes can be divided into internal and external causes (Howe 1909: 44–49; Terzaghi 1960: 88 in Erismann and Abele 2001: 107). This subdivision is based on the shearing properties of material stating that internal causes are those which lead to a sliding mechanism without external influence and without a change in surface conditions. The external causes lead to an increase in shearing stresses. The example of gravitational forces, which can be grouped neither to internal nor external causes, might indicate the issue of applying this classification in practices. However it is a first approach of suggesting a classification scheme. It seems more suitable to divide the promoters of landslides in causes; meaning factors which induce landslides over a certain period of time and one trigger, which instantaneously leads to failure (Varnes 1978 in Cruden and Varnes 1996: 76).

A classification of causes is given by four basic parameters and their interaction (Erismann and Abele 2001: 109):

- The slope angle β of the underlying ground surface
- The coefficient of friction μ between the potential mass of failure and the ground
- The cohesion c between the critical mass and the surrounding rock mass
- The external forces (for example the hydraulic pressure in rock joints)

Further classification suggests the categorisation into the following four classes (Cruden and Varnes 1996: 70): geological causes, morphological causes, physical causes and human causes. For the process of rockfalls a few examples for each group are mentioned as:

- Geological causes: adversely oriented discontinuities, weathered materials
- Morphological causes: toe-erosion; steep valley flanks
- Physical causes: thawing, intense rainfall, freeze-and-thaw weathering,
- Human causes: excavation of the slope toe

Both classification systems interact with each other since for example the geological causes like weathered materials would affect the coefficient of friction as well as the parameter of cohesion.

2.2.2 *Critical Fracture Propagation (Progressive Failure) and Triggers*

Continuous Crack Propagation

For the failure of a potential rock mass a growth of cracks must occur so that a coherent failure plane can be formed. In rocks this implies the transition from rock bridges to roughness contacts, which are separated by cracks. A rock bridge is defined as a segment of intact rock separating co-planar or non-coplanar discontinuities from each other (Kemeny 2005: 36).

If a critical mass fails due to crack propagation over a certain period of time two terms can be classified referring to the location of stress attack (Erismann and Abele 2001: 121 ff.): if one single location or rock bond is affected by several load cycles the process would be named as "fatigue"; if many locations at a failure surface are affected by changing stress concentration this would be called as "progressive failure". So the failure of a single rock bridge could be considered as fatigue in the general context of progressive failure. The term "fatigue" was introduced giving the following detailed definition (Visser 1980 in Erismann and Abele 2001: 120): "… fatigue is the property of material to fail after many repetitions of a deforming stress, which by itself is not high enough to cause failure." This definition implies that a failure occurs not until several load cycles have occurred. The following formula was set as a general idea for fatigue crack propagation (Paris 1962 in Erismann and Abele 2001: 120):

$$\frac{dc}{dN} = (C\Delta K)^n \tag{2.3}$$

where dc represents the crack growth per load cycle (with N as the cycle number), ΔK stands for the variation of stress concentration and C and n are constants, determined experimentally. The relation shows that the crack growth increases due to increasing stress concentrations. Or the other way round, if the cracks grow further, the stress concentration will increase during the load cycles. Since the driving force of a critical mass can be assumed as more or less constant due to a direct relation to gravity, the crack growth is increasing in an accelerated pace.

Assuming a failure surface of several square meters, repeating load cycles lead to failure of one or more rock bridges. The consequence would be a redistribution of stresses among the remaining rock bridges and asperity contacts. With every time one or more intact rock contacts fail, the induced stresses affect the remaining intact contacts over a certain period of time until the critical mass fails as a whole. The redistribution of stresses affects interlocking asperity contacts in the same way as intact rock bridges. Interlocked roughness contacts could also be sheared off due to increasing shear stresses. A sketch of the transformation from rockbridges to roughness contacts is given in Fig. 2.5.

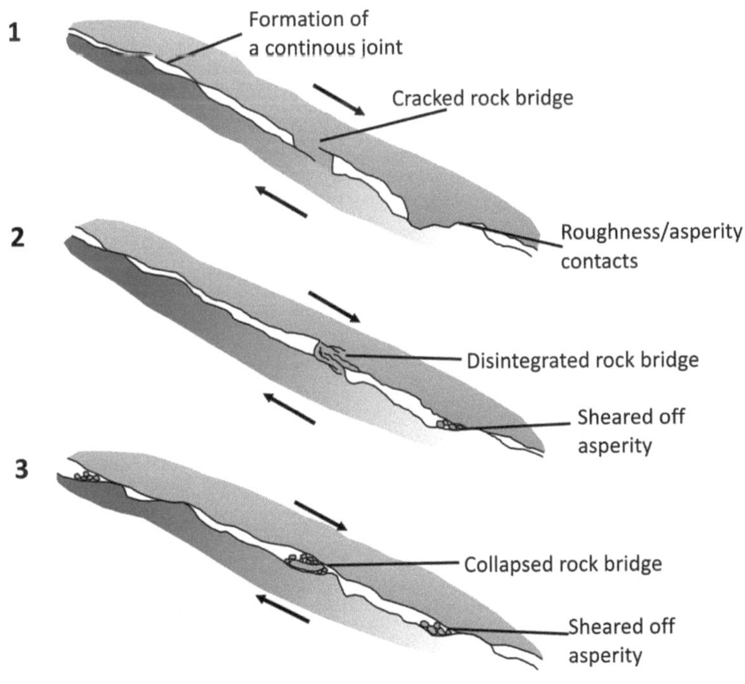

Fig. 2.5 Stages in transformation of rock bridges to asperity rock-rock contacts at a failure surface

The formation of cracks and at larger scale discontinuities has to be embedded in the context of stress regime affecting the different part of a critical rock mass. There are three types of discrete stress distributions which are classified due to the orientation of the stress vector in relation to the surface of a potential joint (Erismann and Abele 2001: 116 ff.):

- Mode I: the tensile stress vector is oriented perpendicular to the joint surface. This mode is named according to the mouth of an alligator.
- Mode II: The shearing stress vectors are oriented parallel to the shearing plane but perpendicular to the margin surface of a potential block. This mode is named referring to the mouth attitude of a bulldog.
- Mode III: The shearing stress vectors are oriented parallel to the shearing plane and also parallel to the edge of the block. This mode is named according to the mouth attitude of a cow.

The three modes of stress distribution can occur considering a case of planar failure on an inclined failure surface (Fig. 2.6). At the detachment plane tensile stress is dominating, which leads to crack opening from top to bottom. At the

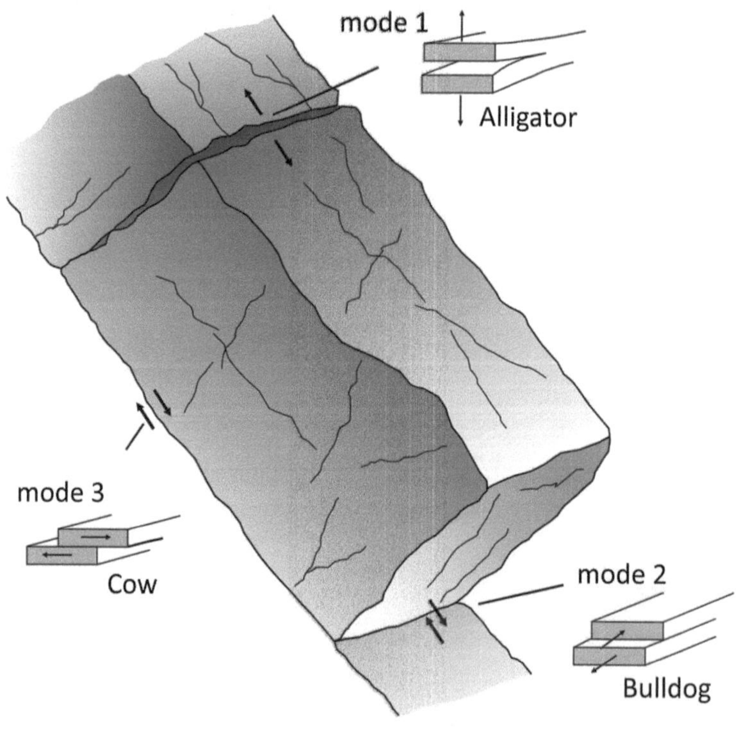

Fig. 2.6 The mechanical modes of failure named according to animal's mouth attitudes (after Erismann and Abele 2001: 116) illustrated using the example of planar rock mass failure

bottom of the block mode II can be assumed, whereas the shearing vectors at the lateral edges of the block can be grouped to mode III.

Trigger Mechanisms

"The trigger of an event is the last cause which puts an end to the balance between driving and retaining forces and thus sets the mass moving."(Erismann and Abele 2001: 108)

In many cases triggers can be seen as external drivers, for example: intense rainfalls, rapid snowmelt, water-level changes, earthquake shaking and volcanic eruption (Wieczorek 1996: 76 ff.). For rockfalls especially the thawing of ice and ice segregation can play a certain role as triggers (Krautblatter et al. 2013). A trigger leads to a strong decrease in the retaining forces so that failure occurs as the consequence.

An issue with the assignment of triggers is that in several cases an exact trigger cannot be determined. If the in Sect. 2.2.1 mentioned causes affect a potential mass over a critical period of time, this could lead to material fatigue of the rock mass, which consequently leads to failure. The consequence is a progressive failure of a critical mass.

Types of Detachment: Topple, Slide (Block and Wedge Sliding), Rock Slumping

The detachment process of rock masses depends on several factors: Joint systems and their orientation, spacing of the joints and inclination of the slope/rock face. The mentioned factors influence the released volume, the block shape and especially the detachment process (Fig. 2.7). The general detachment mechanisms for rock slopes could be divided into the following groups: planar sliding, wedge

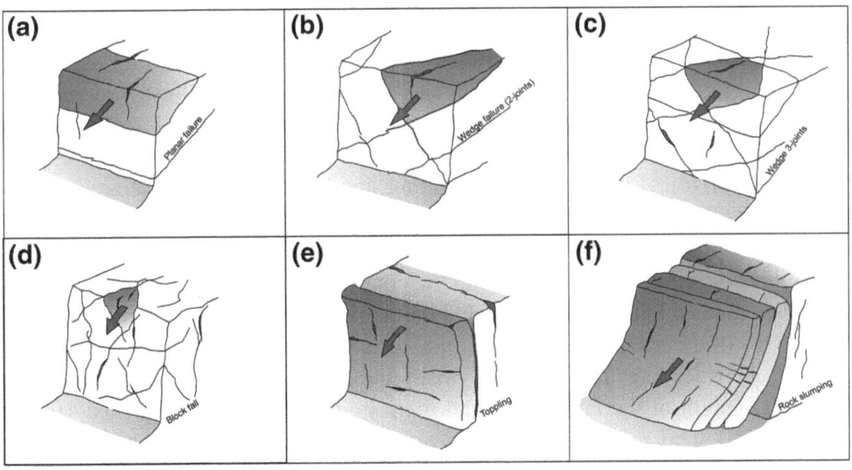

Fig. 2.7 Examples of rock mass failures referring to their mode of release: Planar sliding (**a**), wedge sliding (**b** and **c**), free fall (**d**), toppling (**e**) and rock slumping (**f**)

sliding, toppling, rock slumping, buckling and free fall (Whalley 1984; Goodman and Kieffer 2000).

The planar failure occurs on one single plane of failure due to undercutting of a rock slope by one joint system (Fig. 2.7a). If two joints are intersecting in an adverse way and the line of intersection is inclined parallel to the slope this could result in wedge sliding. The wedge could be defined by 2 or more joints (Fig. 2.7b, c). If a joint system draws through a rock mass totally undefined, blocks of an undefined shape could be released and descend the slope (Fig. 2.7d). In case of a very steep and parallel to the slope inclined joint system toppling can occur. In case of toppling a rock slab fails in a forward rotational movement, since the centre of gravity is exposed over the edge of the rock slab. The tension cracks between the rock slabs show a V-shaped form (Fig. 2.7e). Due to a missing toe support rock slabs are released in a backward rotation, which leads to rock slumping. The rock slabs are separated by A-shaped tension cracks (Fig. 2.7f).

The spacing of a joint system affects the volume of the released mass due to the variation in length and width.

2.2.3 *Rockfall Run-Out Modelling Approaches*

A Short History of Rockfall Modelling

In this paragraph a short overview of the history of rockfall modelling should be given, whereas the detailed description of the different modelling types is presented in Sect. 2.2.3.

The need for run-out assessment of rockfalls goes back to the 1930s (Heim 1932). Heim was the first author who described the behaviour of blocks during free fall and the subsequent modes of motion through the travelling process. The complexity of the impact of a falling block/boulder on the slope surface including a potential degree of fragmentation and the dispersion of trajectories is illustrated in detail (Heim 1932, 64 ff.). The first empirical approach of run-out modelling, namely the method of the "Fahrböschung" (Fig. 2.8), suggests the measuring of the angle between a horizontal line and the connecting line between the release area and the farthest travelled block at the run-out area (Evans and Hungr 1993; Heim 1932). A similar approach is given by the "minimum shadow angle"-principle including the angle between a horizontal line and the connecting line between the toe point of the source area and the farthest travelled block of a rockfall mass (Evans and Hungr 1993). One of the enhanced empirical models is the run-out-ratio model, describing the ratio between the horizontal length of the run-out distance and a combined horizontal length of the talus slope and the rock face (McClung and Lied 1987 in Dorren 2003). This model is based on a run-out approach for avalanche modelling.

The next step in the development of rockfall models are the process based models reducing the rockfall process to a 2D section, where the lateral dispersion is not considered. The modes of motion during the falling process were combined of flying and collision stages, where the flying stages were modelled as parabolic

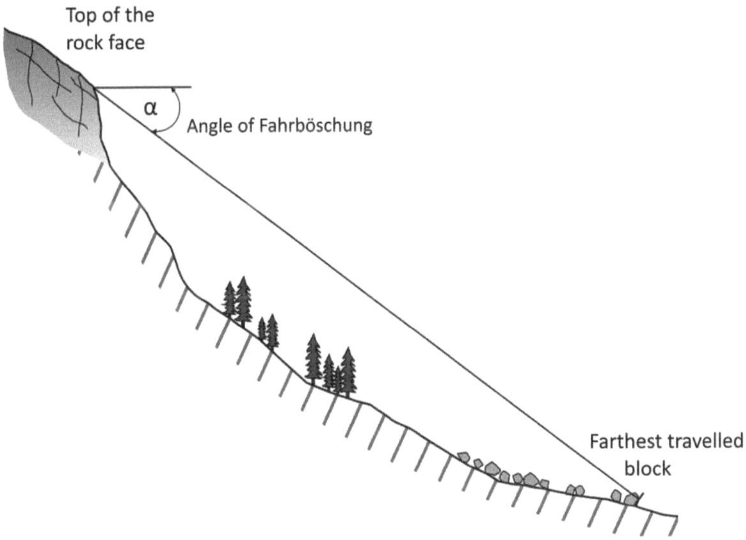

Fig. 2.8 Illustration of the Fahrböschung as an empirical approach for run-out assessment

throws with velocities in x and y direction, where the acceleration of the block is based on gravity (Dorren 2003). There are plenty of process based models describing the rockfall process in a 2D limited space as mentioned above (Azzoni and Freitas 1995; Hungr and Evans 1988 in Dorren 2003; Statham 1976).

The 3D rockfall models are based on a GIS approach or a combined GIS process based approach. The GIS based models are composed of three modules: the determination of the source area, the calculation of falling tracks and the calculation of the run-out (Hegg and Kienholz 1995 in Dorren 2003). The evolution of the GIS based models reaches from models which simulate the block sliding over the slopes surface to models simulating an initial free fall and the subsequent bouncing and rolling. The GIS-process based models, namely Rockyfor 3D can be used for rockfall calculations at regional scale (Dorren 2003).

The Scope of Rockfall Modelling

The applied purpose of rockfall modelling is the improvement of hazard assessment, including the run-out prediction of possible events as well as the design of mitigation measures. Apart from any certain code the most important outcomes are:

- The run-out distance: to determine whether critical blocks will reach endangered infrastructure or not.
- The velocity and kinetic energy: as linked dimensions to design the load capacity of mitigation measures.
- The jumping heights: to adapt the height of fences or walls to the critical jumping height of descending blocks.

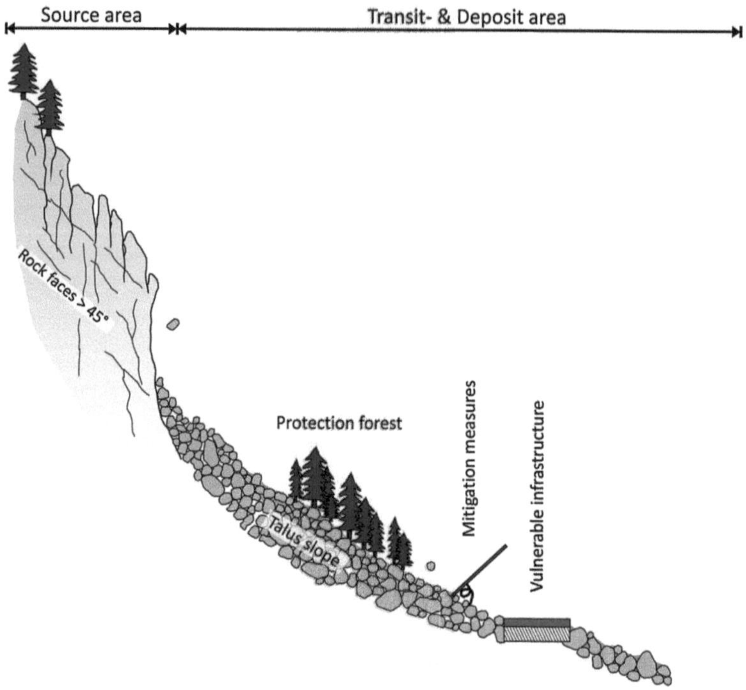

Fig. 2.9 General sketch of the rockfall process areas: the source area, the transit- and deposit area. Protection forest as well as fences or walls are retaining measures at endangered slopes. The critical inclination angle is assumed as 40° for the rockfall source areas (after Dorren 2003)

The crucial input parameters can be categorized according to the process areas of a rockfall slope (Fig. 2.9). For the source area the block parameters like the dimensions, the shape and the material density have to be defined. The transit and deposit area need to be defined by roughness and damping parameters of the ground to assess the decelerating forces for the block (Dorren 2012; Krummenacher et al. 2005; Guzetti et al. 2002). Since natural obstacles like protection forest have a decelerating effect on descending blocks, the forest stand is an important parameter which is implemented into the state of the art 3D rockfall codes (Dorren 2012; Bartelt et al. 2013 Krummenacher et al. 2005).

Modes of Interaction During the Rockfall Process and Modelling Approaches
The rockfall process is characterized by the following stages of motion: the detachment-, the transition- and the stopping or deposit phase. The detachment often leads over to a free fall stage, where the duration of the free fall depends on the inclination angle of the terrain. The transition stage is composed of flying and collision stages, which is called bouncing (Dorren 2003). As the total kinetic energy is decreasing, the mode of motion could change from bouncing to rolling until the block finally stops. This point is reached, when the kinetic energy loss due to

collisions with the underground, forest stems or obstacles is so high that the block is forced to stop.

Since the current thesis deals with 3D rockfall modelling, the state of the art in terms of modelling refers to the state of the art 3D-codes.

Interaction Block—Underground

The interaction of a descending block with the underground material is an important influencing factor of the blocks energy loss. The state of the art 3D rockfall codes nevertheless differ in the approaches of describing the contact between boulder and underground.

One of the most applied approaches is to characterize the process of energy loss due to slope contacts by two main parameter sets (Dorren 2012; Guzetti et al. 2002; Krummenacher et al. 2005):

- The damping: represented by the normal and tangential coefficient of restitution (R_N and R_T). The coefficient of restitution is defined as:

$$R = \frac{V_{ac}}{V_{bc}} \tag{2.4}$$

  where v_{ac} is the speed after collision and v_{bc} is the speed before collision.
- The roughness: represented by undulation of the underground material properties

The coefficients of restitution are important for the calculation of the penetration depth of the block into the ground material (R_N) and the calculation of the velocity after the rebound (R_T). Their integration into the algorithm depends on the code (Bourrier et al. 2009; Dorren 2012; Guzetti et al. 2002).

The slope roughness can be described by the diameter of blocks building a talus slope (Krummenacher et al. 2005). An enhanced way of determining the "true" effect of the slope roughness to descending blocks would be the assessment of the mean obstacle height (MOH). The MOH mirrors the height of an obstacle considered in the descending line of the falling block (Dorren 2012; Fig. 2.10) Apart from any approach of describing the roughness, the slope roughness has got an evident effect on blocks with a small diameter compared to the amplitudes of roughness. The effect of slope roughness decreases if the block diameter increases in comparison to the amplitudes of slope roughness.

In most codes where the rebound approach is based on the coefficients of restitution, a complex or natural block shape is not considered. The Code STONE 3D considers the block as a lumped mass, implying that the mass is concentrated in one central point (Guzetti et al. 2002: 1082). The code Rockyfor3D provides the possibility to enter a block shape. Nevertheless during the contact block—slope, the boulder is considered to be a sphere with a diameter composed of the mean of the two longest block axes (Fig. 2.11; Dorren 2012: 13 ff.).

One of the most actual rebound approaches provides the possibility to take the shape of a descending rock fully into account (Fig. 2.11). The physical modelling

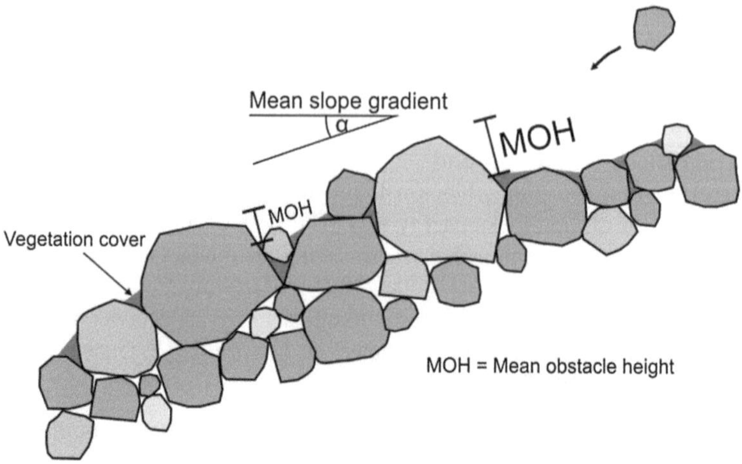

Fig. 2.10 Illustration of the Mean obstacle height (MOH) (after Dorren 2012), which implies the height of an obstacle in falling direction of a descending rock

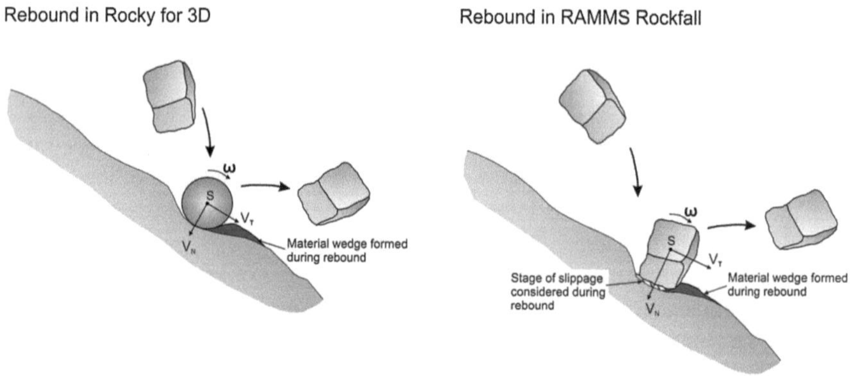

Fig. 2.11 *Left* Rebound analysis in Rockyfor 3D: The rock is considered as a sphere during rebound, with its centre of mass at the centre point of the sphere. *Right* Rebound analysis in RAMMS::Rockfall: the rock shape is taken into account during rebound, the block stick-slides over the terrain before take-off

approach is based on the non-smooth contact dynamics method and was implemented into the software RAMMS::Rockfall. The descending block is modelled as a rigid complex shaped body, which is characterized by its centre of mass and inertia vector (Leine et al. 2013; Bartelt et al. 2013).

The following important issues in terms of rock—slope interaction are considered in this model:

- The three dimensional motion of a 3D shaped rigid body in space (with three degrees of freedom for translational motion and three degrees of freedom in terms of rotation)

- The consideration of a 3D shaped block during the slope contact (the models described above consider the block as a sphere during the block/terrain interaction)
- The stage of slippage during the rebound process

In the code RAMMS::Rockfall the terrain geometry is described by a high resolved digital terrain model (DTM). The terrain materials properties are defined referring to the Coulomb-Friction law, where the terrain material is determined by the friction coefficient and the coefficients of restitution (Bartelt et al. 2013).

The Interaction Block—Forest Stand

The protective effect of mountain forests was in the focus of research over the last years (Berger et al. 2002; Jahn 1988; Dorren et al. 2007; Stoffel et al. 2006). The presence of forest stand on alpine hill slopes leads to a decrease in velocities, rebound height and kinetic energy. Consequently the run-out distance will be reduced due to forest stand (Dorren et al. 2005). Studies on the required species of a protective forest show, that broadleaved trees would have the more robust properties against rockfall than coniferous trees (Stokes 2006: 84; Dorren and Berger 2005: 69). Nevertheless a mix of both tree species would be most effective since coniferous trees are more effective in preventing avalanche damage (Stokes 2006). Apart from the tree species it seems to be more effective to provide a forest stand with a large number of trees instead of a decreasing number of trees with a large diameter at breast height.

To directly compare the effect of protection forest on rockfall hazard, real size experiments were carried out on a slope providing an equivalent geological set up, but with and without forest stand (Dorren et al. 2006).

The forest stand can quantitatively be described by the following parameters:

- Tree species
- The diameter at breast height (DBH)
- The stock density, which can be assessed via:
  - The number of tree stems per hectare
  - The mean tree free distance (MTFD) (Krummenacher et al. 2005)

The drag effect of the forest stand can be implemented in rockfall codes using different approaches. A detailed approach of describing forest stand is to define homogenous areas for the forest stand by mapping. For each area the number of tree stems per hectare and the percent of coniferous trees have to be determined. To assess the resistance against stem breakage, the diameter at breast height is specified as a mean value with the associated standard deviation. The described parameters are required for the software package Rocky for 3D. In this code the forest stand is defined pixel based, meaning that the number of trees is randomly defined for each pixel with the associated DBH (Dorren 2012).

Another approach is that the forest stand is simply defined as an additional drag force depending on the tree height. Every time the block descends the mean forest height the drag force is applied. This approach is applied in the code RAMMS:: Rockfall (Bartelt et al. 2013).

2.3 Key Questions and Research Gap

The frame of this research work is set by the topic of rockfall hazard assessment. The described state of the art demonstrates the stress field between rockfall run-out modelling and the mechanics of failure preparation in advance of release.

The submitted PhD thesis aims to link the field of detachment and failure mechanics to the field of rockfall run-out analysis by addressing the following key-questions:

1. Detachment mechanics at the rockfall source area:
   (a) How can we achieve reconnaissance about the failure mechanics of mid-magnitude blocks?
   (b) How can we analyse the mechanical behaviour of a critical block by analysing the limit equilibrium stage?
   (c) How can we provide accurate information on the degree of potential fragmentation as input parameter for rockfall modelling?
2. Quantitative magnitude assessment and detachment processes:
   (a) How can we validate assumed detachment processes by scanline and kinematic analyses?
   (b) How can we provide a structured way of evaluating rock volumes in a quantitative way?
   (c) How can we transfer our knowledge from block recording to reasonable information in terms of rockfall modelling?
3. Run-out modelling
   (a) How does an increased slope roughness affect the rockfall run-out?
   (b) How do areas of windfall affect the rockfall run-out in terms of increased slope roughness?
   (c) How can we consider potential degrees of fragmentation in detailed 3D modelling studies and which assumptions do we have to take?

References

Aydin A, Basu A (2005) The schmidt hammer in rock material charcterisation. Eng Geol 81:1–14
Azzoni A, Freitas MH (1995) Experimentally gained parameters, decisive for rockfall analysis. Rock Mech Rock Eng 28(2):111–124
Azzoni A, Barbera GL, Zaninetti A (1995) Analysis and prediction of rockfalls using a mathematical model. Int J Rock Mech Min 32:709–24
Bartelt P, Buehler Y, Christen M, Dreier L, Gerber W, Glover J, Schneider M, Glocker C, Leine R, Schreizer A (2013) RAMMS. User manual v1.5 Rockfall.-83p. (pdf-version)
Barton N, Choubey V (1977) The shear strength of rock joints in theory and practice. Rock Mech 10:1–54
Barton NR (1973) A review of a new shear strength criterion of rock joints. Eng Geol 7:287–332
Barton N (1976) Rock mechanics review—the shear strength of rock and rock joints. Int J Rock Mech Min Sci Geomech Abstr 13:255–279
Berger F, Quetel C, Dorren LKA (2002) Forest a natural protection mean against rockfalls, but with which efficiency? In: Interpraevent 2002-Conference Proceedings, vol 2, 815–826
Bieniawski ZT (1976) Rock mass classification in rock engineering. In: proceedings of the Symposium on Exploration for rock engineering, vol 1, 97–106
Bourrier F, Dorren LKA, Nicot F, Berger F, Darve F (2009) Toward objective rockfall trajectory simulation using a stochastic impact model. Geomorphology 110:68–79
Cruden DM, Varnes DJ (1996) Landslide types and processes. In: Turner AK, Schuster RL (eds) [Hrsg.]: landslides investigation and mitigation, Special Report 247. National Academy Press, Washington, D.C, pp 38–75
Deere DU, Merrit AH, Coon RF (1969) Engineering classification of in-situ rock. Technichal Report No. AFWL-TR-68–144. University of Illinois, Illinois, 293 p. (pdf-version)
Dorren LAK (2003) A review of rockfall mechanics and modelling approaches. Prog Phys Geogr 27(1):69–87
Dorren LAK, Berger F (2005) Stem breakage of trees and energy dissipation during rockfall impacts. Tree Physiol 26:63–71
Dorren LAK, Berger F, Hir C, Mermin E, Tardif P (2005) Mechanisms, effects and management implications of rockfall in forests. For Ecol Manage 215:183–195
Dorren LAK, Berger F, Hir C, Mermin E, Tardif P (2006) Results of real size rockfall experiments on forested and non-forested slopes. Disaster Mitig Debris Flows Slope Failures Landslides 223–228
Dorren L, Berger F, Jonsson M, Krautblatter M, Mölk M, Stoffel M, Wehrli A (2007) State of the art in rockfall forest interactions. Schweiz Z Forstwes 158(6):128–141
Dorren LKA (2012) Rocky for 3D (v5.0) revealed—Transparent description of the complete 3D rockfall model. ecorisQ paper (www.ecorisq.org) 31 p
Einstein HH, Veneziano D, Baecher GB, O'Reilly KJ (1983) The effect of discontinuity persistence on rockslope stability. Int J Rock Mech Min Sci Geomech Abstr 20(5):227–236
Erismann TH, Abele E (2001) Dynamics of rockslides and rockfalls. Springer, Berlin, 315 p
Evans SG, Hungr O (1993) The assessment of rockfall hazard at the base of talus alopes. Can Geotech J 30:620–636
Goldstein M, Goosev B, Pyrogovsky N, Tulinov R, Turovskaya A Investigation of mechanical properties of cracked rock. In: Proceedings of 1st Congress International Society for Rock Mechanics, vol 1. Lisbon, pp 521–524
Goodman RE, Kieffer DS (2000) The behavior of rock in slopes. J Geotech Geoenviron Eng, 675–684 (pdf-version)
Goudie AS (2006) The schmidt hammer in geomorphological research. Prog Phys Geogr 30 (6):703–718
Guzetti F, Crosta G, Detti R, Agliardi F (2002) STONE 3D—a computer program for the three dimensional simulation of rock-falls. Comput Geosci 28:1079–1093

Hack R, Price D, Rengers N (2002) A new approach to rock slope stability—a probability classification (SSPC). Bull Eng Geol Environ 62:167–184. doi:10.1007/s10064–002-0155-4
Hegg C, Kienholz H (1995) Determiningpaths of gravity-driven slope processes—the'Vector Tree Model'. In: Carrara A, Guzetti F (eds]) Geographic information systems in assessing natural hazards. Dordrecht, pp 79–92
Heim A (1932) Bergsturz und Menschenleben. Fretz & Wasmuth Verlag, Zürich, 117 p
Hoek E (2000) Practical rock engineering, 313 p. (pdf-version, www.rocscience.com)
Howe E (1909) Landslides in the San Juan mountains, Colorado. Professional Paper 67, U.S. Geological Survey, Washington, 58 p
Hudson JA, Harrison JP (1997) Engineering rock mechansics—an introduction to the principles. Pergamon, Amsterdam, 444 p
Hudson JA, Priest SD (1979) Discontinuities and rock mass geometry. Int J Rock Mech Min Sci Geomech Abstr 16:339–362
Hungr O, Evans SG (1988) Engineering evaluation of fragmental rockfall hazards. In: Proceedings of fifth international symposium on ladslides, vol 1,pp 685–690
ISRM—International Society for Rock Mechanics (1978) Suggested methods for the quantitative description of discontinuities in rock masses—Commission on standardization of laboratory and field tests. Int J Rock Mech Min Sci Geomech Abstr 15(4):319–368
Jahn J (1988) Entwaldung und Steinschlag. Proceedings of the Conference Interpraevent 1988, vol 2, pp 185–198
Kemeny J (2005) Time dependent drift degradartion due to progressive failure of rock bridges along discontinuities. Int J Rock Mech Min Sci 42:35–46
Kersten RWO (1990) The stress distribution required for fault and joint develop-ment. In: Proceedings of the International Conference on Mechanics of Jointed and Faulted Rock, Balkema, Rotterdam, pp 251–6
Krautblatter M, Funk D, Guenzel F (2013) Why permafrost rocks become unstable: a rock-ice mechanical model in time and space. Earth Surf Processes Land 38(8):876–887
Krummenacher B, Pfeifer R, Tobler D, Keusen HR, Liniger M, Zinggeler A (2005) Modellierung von Stein- und Blockschlag—Berechnung der Trajektorien auf Profilen und im 3-D Raum unter Berücksichtigung von Waldbestand und Hindernissen. 10 S., Zollikofen (pdf-Version)
Leine RI, Schweizer A, Christen M, Glover J, Bartelt P, Gerber W (2013) Simulation of rockfall trajectories with consideration of rock shape. Multibody Syst Dyn 31 p. doi:10.1007/s11044-013-9393-4. Retrieved as pdf-version at Springer Science + Business Media, Dordrecht
McCarroll (1997) A template for calculating rock surface roughness. Earth Surf Processes Land 22:1229–1230
McClung DM, Lied K (1987) Statistical and geometrical definition of snow avalanche run-out. Cold Reg Sci Technol 13:107–119
Norrish NI, Wyllie DC (1996) Rock slope stability analysis. In: Turner AK, Schuster RL (eds) [Hrsg.]: Landslides investigation and mitigation, Special Report 247. National Academy Press, Washington, pp 38–75
Paris PC (1962) The growth of cracks due to variation in load. Doctor's Thesis, University Lehigh, Pennsylvania
Patton FD (1966) Multiple modes of shear failure in rock and related materials. Thesis, Univ. III, 282 p
Pierson LA (1991) The rockfall hazard rating system. Oregon department of transportation (pdf-version)
Price NJ (1966) Fault and joint development in brittle and semi-brittle rock. Oxford, Pergamon
Priest SD (1993) Discontinuity analysis for rock engineering. Chapman & Hall, London, 473 p
Ritchie AM (1963) Evaluation of rockfall and its control. Highw Res Rec 17:13–28
Selby MJ (1982) Controls on the stability and inclinations of hill slopes formed on hardrock. Earth surf land 7:449–467
Statham I (1976) A scree slope rockfall modell. Earth Surf Processes 1:43–62

Stoffel M, Wherli A, Kühne R, Dorren LA, Perret S, Kienholz H (2006) Assessing the protective effect of mountain forests against rockfall using a 3D simulation model. Forest Ecol Manage 225:113–122

Stokes A (2006) Selecting tree species for use in rockfall protection forests. For Snow Landsc Res 80(1):77–86

Terzaghi K (1946) Rock defects and loads on tunnel supports. In: Proctor RV, White Rock TL (eds) Tunneling with steel supports, vol 1, pp 17–99

Terzaghi K (1960) From theory to practice in soil mechanics. Wiley, New York, 425 p

Tse R, Cruden DM (1979) Estimating joint roughness coefficients. Int J Rock Mech Min Sci Geomech Abstr 16:303–307

Varnes (1978) Slope movement, types and processes. In: Special Report 176: Landslides analysis and control, TRB National Research Council, Washington D.C., pp 12–33

Visser WA (1980) Geological Nomenclature. R Geol Min Soc Neth (The Hague Boston London)

Whalley (1984) Rockfalls. In: Brunsden D, Prior DB (1984) Slope instability. Wiley, New York, 620 p

Wieczorek GF (1996) Landslide triggering mechanisms. In: Turner AK, Schuster RL (eds) [Hrsg.]: Landslides Investigation and Mitigation, Special Report 247, National Academy Press, Washington, pp 76–90

Yang ZY, Lo SC, Di CC (2001) Reassessing the joint roughness coefficient (JRC) estimation using (Z2). Rock Mech Rock Eng 34(3):243–251

Chapter 3
Study Site

3.1 Geographical Setting

Our study site extends above the federal road B 305 between the villages Unterjettenberg and Schwarzbachwacht in the Bavarian Alps and is situated approximately 30 km southwest of the city of Salzburg (Fig. 3.1). Figure 3.2 shows an overview of the project site with a view from south to north, where the project site extends at the east (right) valley side. The terrain can be characterized as a steep and densely forested limestone slope with two extensive cliffs in two altitude levels in the middle and upper slope, providing the majority of the rockfall material. The cliff faces dip with an angle of 50°–90° in west direction. The slope below the cliffs is covered with talus material, which could, based on the slope inclination angle of 20°–50°, be released as secondary rockfall material. The rock faces as well as the talus material consist of massive unclear bedded carbonates belonging to the Dachstein-Formation. The cliffs tend to build up steep but mostly not vertical, benched rock faces due to an orthogonal joint system.

3.2 Geological Setting

Since the study site is located in the Bavarian part of the Northern Calcareous Alps, the chapter Geology is limited to the description of the Northern Calcareous Alps as a part of the Oberostalpin.

The Oberostalpin is divided into three geo-tectonical units: the Northern Grauwackenzone, the Drauzug-Zone and the Northern Calcareous Alps. The Northern Calcareous Alps are internally divided into three tectonical units (Hahn 1912: 338 ff.; Tollmann 1976: 47); namely from the bottom to the top: The Bajuvarical, the Tirolic and the Juvavic Units. The Bajuvaric Unit contains the so called Cenoman Nappe, the Allgäu Nappe (Lower Bajuvarikum, Tiefbajuvarikum)

B. Sellmeier, *Quantitative Parameterization and 3D-run-out Modelling of Rockfalls at Steep Limestone Cliffs in the Bavarian Alps*, Springer Theses,
DOI 10.1007/978-3-319-24510-2_3

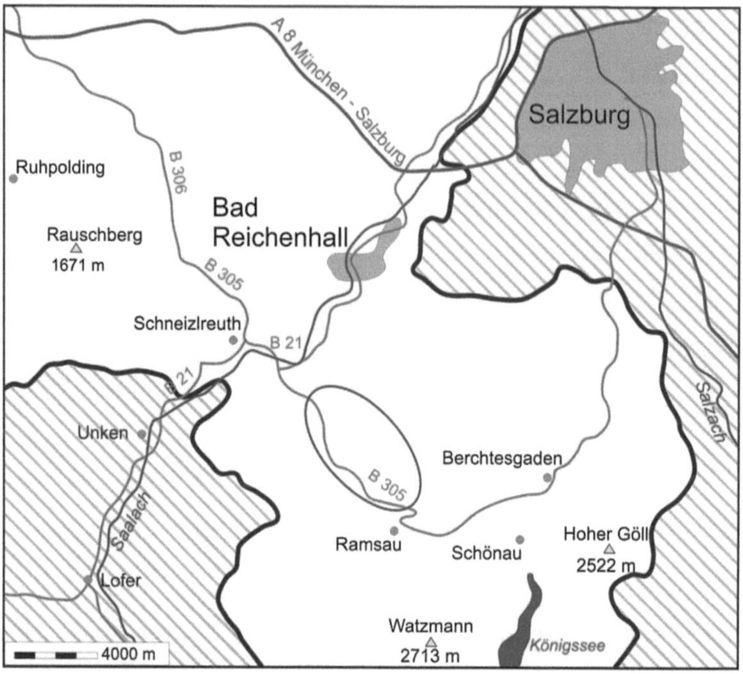

Fig. 3.1 Simplified geographic map of the project site and its vicinity; the study site extends along the federal road B 305 and is marked with a *red oval shape*

Fig. 3.2 Overview of the Schwarzbach-Valley with a view from south to north. The study site extends along the east valley slope. The federal road is situated at the bottom of the densely forested slope

and the Lechtal Nappe (Higher Bajuvarikum, Hochbajuvarikum). The mentioned nappes occur mainly in the western Northern Calcareous Alps, but in total they can be observed from the Rätikon in the West to the Wiener Wald in the East (Gwinner 1971: 238 ff.).

The overlaying tectonic units like the Inntal-Nappe in the west and the Staufen-Höllengebirgs-Nappe in the east, can be embraced in the Tirolicum. The vicinity of the study site is composed of sediments belonging to the Staufen-Höllengebirgs-Nappe (Fig. 3.3).

The tectonic units the building the top at the region around the project site are the Juvavic "Nappes", which can be divided into two structural levels: the lower juvavicum (Tiefjuvavikum) and the Upper Juvavicum (Hochjuvavicum) (Hahn 1912: 338 ff.). The Lower Juvavicum is composed of geological units of the Hallstätter-Facies, whereas the Upper Juvavicum contains large-scale nappes including permo-skyth to tertiary sediments. The Berchtesgadener- and the Dachstein-Nappe are examples for the Upper Juvavic units.

At the region of Berchtesgaden the tectonic structure is composed in the following way from bottom to top: the Tirolic units (Staufen-Höllengebirgsdecke)—the Lower Juvavic Nappes—the Upper Juvavic Nappes. The genesis and the contact between the three units were discussed controversially over the last years. Referring to Tollmann (1976) the contact between both units is a clear tectonic and lithologic one. Actual approaches suggest the Juvavic Nappes to be olistolithes and slumping sequences having slid into the marine basins during orogenesis (Frisch and Gawlick 2003; Gawlick and Frisch 2003). In the current thesis the classification of Hahn (1912: 338 ff.) and Tollmann (1976) is used due to the clear structure.

The following paragraph gives a short overview of the geological setting at the study sites and their vicinity.

The study site "Weißwand Wald" is located in carbonates of the Dachstein-Formation belonging to the Upper Juvavic Unit, namely the Berchtesgadener-or Reiteralm-Nappe (Tollmann 1976). These carbonates could be considered as a part of the noric bedded back reef and lagoon facies (Bögel and Schmidt 1976: 129; Tollmann 1976: 206 ff.), since the rhätic Dachstein-Formation has been eroded (Risch 1993: 34). The carbonates of the Dachstein-Formation at the project site show the following chracteristics: middle to wide standing, eastwards dipping bedding planes, partly substitution of Megalodons by dolomized layers with a transition to Dachstein-Dolomite (Tollmann 1976: 209 ff.). The carbonates show a light grey weathering color and are dark grey to anthracite colored as fresh material. In the top part of the source area stromatholites can be found.

In the north east of the project site the Dachstein-Formation is discordantely overlaid by cretatous sediments (Risch 1993). Due to the minor influence of the cretaceous sediments on the rockfall activity at the study area, these stratigraphic units will not be described in detail.

The project site is surrounded by two tectonic anticlinal folds: the Jettenberg-Anticlinale (fold axis striking northeast-southwest) in the west and the Totenmann-Anticlinale in the east (fold axis striking approximately north-south).

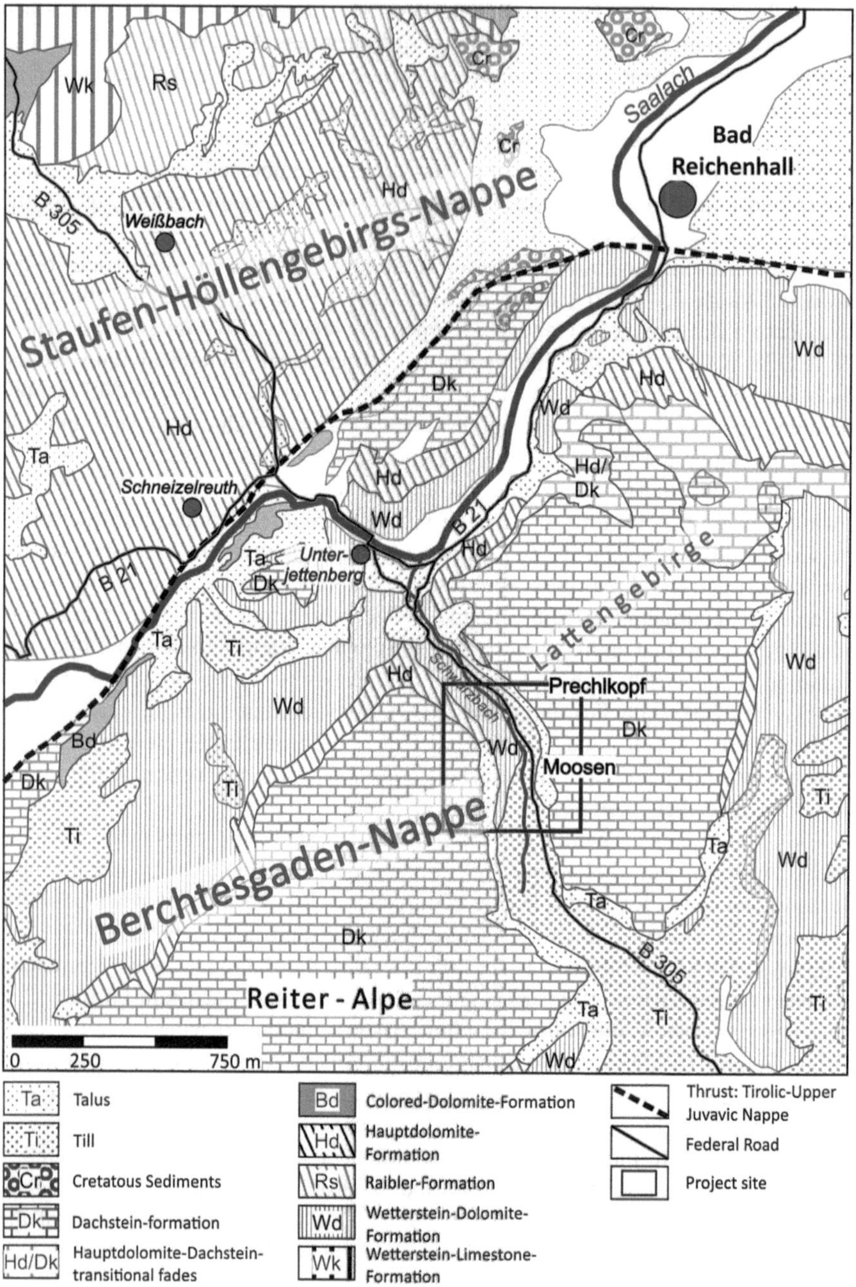

Fig. 3.3 Geologic map of the project site and its vicinity (modified from Sheet 6671 Bad Reichenhall, 1:100,000, Ganss 1978)

References

Bögel H, Schmidt K (1976) Kleine Geologie der Ostalpen. 231 S., Thun (Ott)

Frisch W, Gawlick H-J (2003) The nappe structure of the central Northern Calcareous Alps and its disintegration during Miocene tectonic extrusion—a contribution to understanding the orogenic evolution of the Eastern Alps. Int J Sci (Geol Rundsch) 92:712-727, Berlin, Heidelberg

Ganss O (1978) Geologische Karte von Bayern 1: 100,000, Blatt 667 Bad Reichenhall. München (Bayrisches Geologisches Landesamt)

Gawlick H-J, Frisch W (2003) The Middle to Late Jurassic carbonate clastic radiolaritic flysch sediments in the Northern Calcareous Alps: sedimentology, basin evolution, and tectonics—an overview. N Jb Geol Paläont Abh 230(2/3):163-213, Stuttgart

Gwinner MP (1971) Geologie der Alpen. Stratigraphie, Paläogeografie, Tektonik. 477 S., Stuttgart (Schweizerbart'sche Verlagsbuchhandlung)

Hahn FF (1912) Versuch der Gliederung der Austroalpinen Masse westlich der österreichischen Traun. Verh Geol R.-A 63:337–344

Risch (1993) erläuterungen zur geologischen Karte 1: 25000 Berchtesgaden West Blatt Nr. 8343. Bayrisches Geologisches Landesamt

Tollmann A (1976) Analyse des Klassischen Nordalpinen Mesozoikums. Stratigraphie, Fauna und Fazies der Nördlichen Kalkalpen. 580 S., Wien (Franz Deuticke)

Chapter 4
Methodology

4.1 Case Study: Potential Planar Rock-Slide at the Wachterl-Horn

4.1.1 Field Investigation of the Planar Rockslide

Block Dimensions and Joint Persistence

The block dimensions were assessed using a measuring tape and a folding rule, partly by abseiling. The mapping of the block and its surrounding included consideration of the degree of fragmentation of the block itself and of the surrounding rock mass. Due to the subsequent stability analysis of the whole block the joint persistence was recorded in detail from the out- and inside. Based on the field data a 3D-visualization of the block was developed, showing the dominating joint sets influencing a possible fragmentation scenario. Due to the high inclination angle of the terrain and the limited accessibility it was not possible to accomplish photogrammetric or laser scanning analysis. Figure 4.1 shows the block subjected to planar failure at the source area of the study site.

Recording of the Failure Surface Dimensions

Since the considered block provides the unique occasion of an accessible failure surface, it was possible to perform a detailed investigation of the detachment surface (Fig. 4.2). The fracture roughness was assessed qualitatively for the failure surface in total (ISRM 1978). Further the quantitative roughness parameters were determined by measuring the joint roughness coefficient (JRC) (Barton and Choubey 1977; Sect. 4.1.2).

The dimension of the failure surface underneath the block was mapped in a cm-level of detail (1:50) using a measuring tape and a folding rule. To obtain perception about the opening width and the extension of the cave 4 cross sections equally distributed along the failure surface were recorded. The blocks locked at the

B. Sellmeier, *Quantitative Parameterization and 3D-run-out Modelling of Rockfalls at Steep Limestone Cliffs in the Bavarian Alps*, Springer Theses,
DOI 10.1007/978-3-319-24510-2_4

Fig. 4.1 Block at the release area, a typical example for planar failure. The block volume is approximately 200 m^3

failure surface were localized and mapped. The boundary of the failure cave was categorized in terms of direct rock-rock contacts (rock bridges) and apertured failure surface filled with loosened block material.

4.1.2 Recording of the Shear Parameters at the Failure Surface

To yield reconnaissance about the shear parameters we performed measurements of the joint wall compressive strength (JCS) and the quantitative determination of the joint roughness coefficient (JRC) (Barton and Choubey 1977; ISRM 1978).

The joint compressive strength (JCS) was recorded in situ at the detachment surface as well as at the block-bedrock contacts in the "detachment cave" using a Schmidt Hammer (N-Type) (Barton and Choubey 1977; Woszidlo 1989). We accomplished 200 clustered tests (10 single tests per series) distributed equally over the failure surface, with test areas measuring approximately 10 by 10 cm. The test areas were carefully selected to avoid the influence of surface effects like little

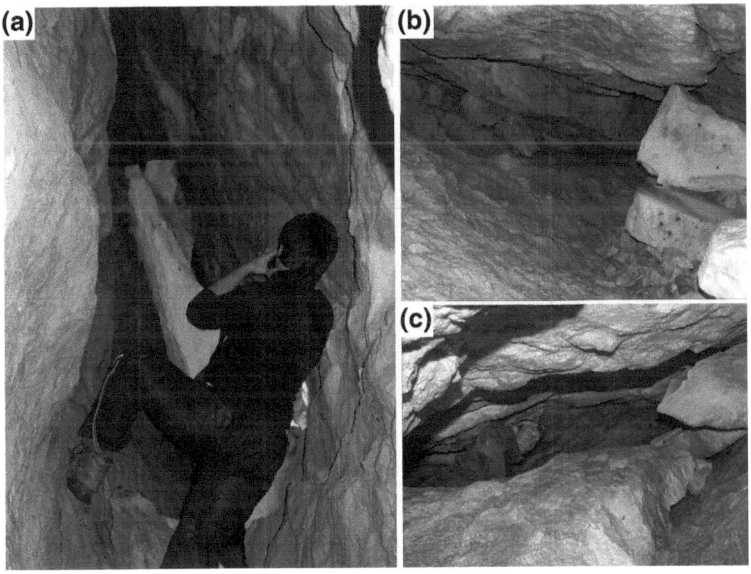

Fig. 4.2 Investigation of the failure surface: a View from north to south; the maximum aperture width is about 1.2 m, the dimensions about 10 m across the slope and max. 6 m upslope. (**b**, **c**) Blocks and rock slabs locked underneath the block

nappes or crusts on the carbonate rock surface. Each of the 20 JCS test series is represented by the arithmetic average of the ten single tests. The results of the in situ-strength testing at the failure surface were complemented by uniaxial compressive strength tests in laboratory experiments (Sect. 4.1.3).

The fracture roughness (joint roughness coefficient, JRC) was recorded along three cross sections in dip direction of the failure plane, where one was situated most in the north, one in the middle part and one in the south part of the failure surface. The surface roughness was quantified using a profile gauge of a length of 260 mm (Fig. 4.3). Two cross sections were 3.12 m (12 gauges) long and one measured 2.60 m (10 gauges). In contrast to the recommendation of McCarroll and Nesje (1996) in McCarroll (1997) we evaluated all recorded single profiles, not only four of them, for each cross section. All JRC profiles were evaluated considering the measuring intervals length of 5, 10 and 20 mm to analyze the influence of the recording interval on the JRC results (McCarroll 1997). The JRC was determined based on the measuring results using the strain gauges referring to the following correlation of the JRC and Z_2 (Yang et al. 2001):

$$\mathrm{JRC} = 32.69 + 32.98\log_{10}Z_2 \tag{4.1}$$

The first derivation (Z_2) of the amplitude meaning the gradient was computed by (Tse and Cruden 1979; Yang et al. 2001):

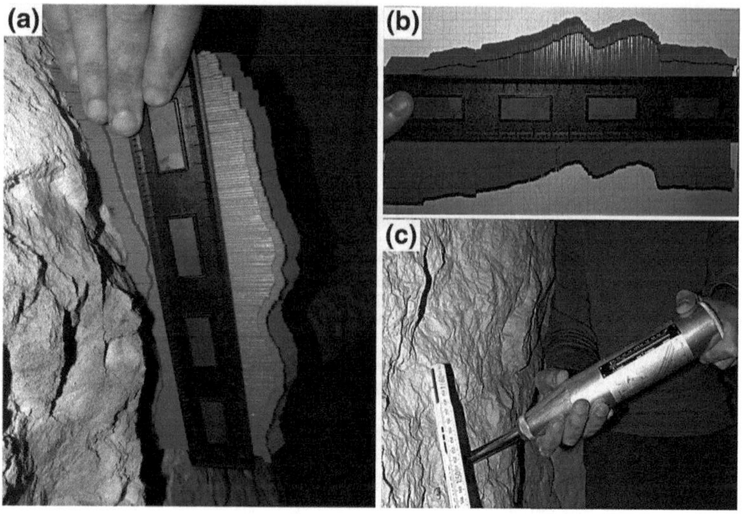

Fig. 4.3 Recording of the shear parameters at the failure surface underneath the block: **a** Joint roughness coefficient (JRC) recording using a fault gauge of 260 mm length. **b** Assessment of the amplitude to length relation using a sheet in mm-scale. **c** Schmidt Hammer Testing at the failure surface

$$Z_2 = \left[\frac{1}{M(Dx)^2} \sum_{i=1}^{M} (y_{i+1} - y_i)^2 \right]^{1/2} \tag{4.2}$$

where M is the number of intervals and Dx refers to the length of the recording interval.

This correlation was recomputed by Yang et al. (2001) based on the work of Tse and Cruden (1979), enhancing an inconsistency in terms of scaling in the former approach.

To finally determine the JRC for each cross section downslope the failure surface, we computed the arithmetic average of the single profiles along each cross section.

4.1.3 Uniaxial Compressive Strength Tests

The results of the Schmidt Hammer testing at the failure surface were validated in the laboratory performing uniaxial compressive strength (UCS) tests. The uniaxial compressive strength tests were performed using cylindrical rock specimen with unobstructed lateral strain (DGGT 2004; ISRM 1978; Thuro et al. 2001). The test results include the determination of the uniaxial compressive strength, the modulus of deformation (V) and the recording of the stress-strain curve (σ-ε-curve).

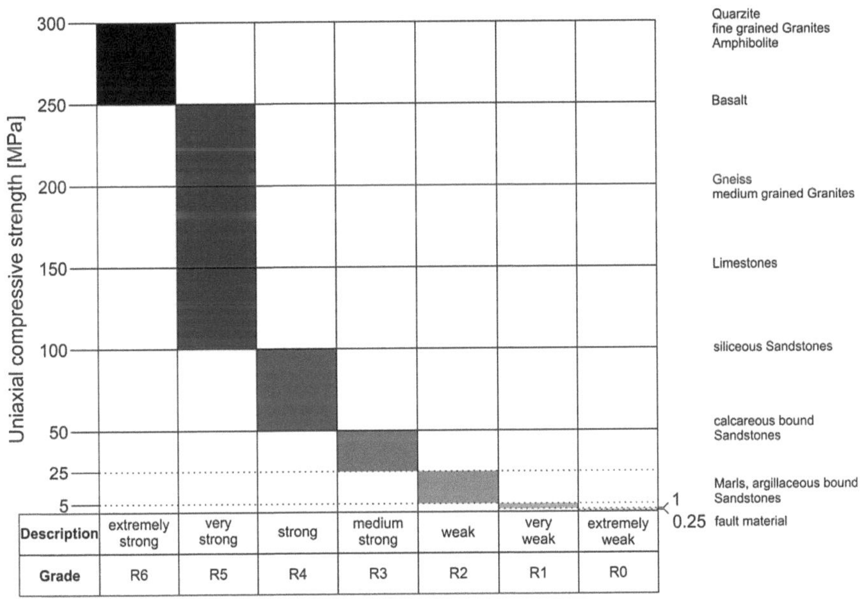

Fig. 4.4 Uniaxial compressive strength classification of hard rocks, including examples (ISRM 1978, modified referring to Thuro 1996)

The specimens were drilled out of four rock samples taken at the Weißwand-project site. The length to diameter proportion of the samples was 2:1, in total 10 carbonate samples of the Dachstein-Formation were tested. Before testing, the dimensions and the mass of all samples were determined. The UCS-tests were performed using a TONI Norm compression test unit, corresponding to class 1 for testing units referring to DIN 51220. The strain in vertical direction was recorded via three inductive displacement transducers. The results were evaluated using the software Test expert (Roell).

All samples were stressed with a load deformation rate of 0.06 mm/min to the point where no rest-shearing resistance existed.

The test evaluation was performed referring to the classification of the ISRM (1978, Fig. 4.4). The uniaxial compressive strength is therefore determined by:

$$\sigma = \frac{F}{A} \tag{4.3}$$

where F is the force of failure, A is the cross-section area of the cylindrical sample.

The strain-modulus (V) was determined in the linear-elastic, pre-failure section of the stress strain curve.

4.1.4 Determination of Density

The materials dry bulk density was determined by floating method (Tauchwägung) under water saturated conditions in the laboratory (DIN 52102 DIN EN 1097-6). Therefor 5 samples of Dachstein carbonate were weighed in pit moist state. Afterwards the samples were stored in a water basin for 24 h. The water saturated samples have been weighed once and second under upwelling conditions. Lastly the samples were dried over 24 h and the weight of the dry samples was recorded again.

4.1.5 Limit Equilibrium Analysis

In a first approach the acting forces at the critical block were determined according to the mechanical approach for a block on an inclined plane (Fig. 4.5). The formulas are included, where c represents the cohesion, φ corresponds to the basic friction angle and A is the total contact area underneath the critical block. The forces are

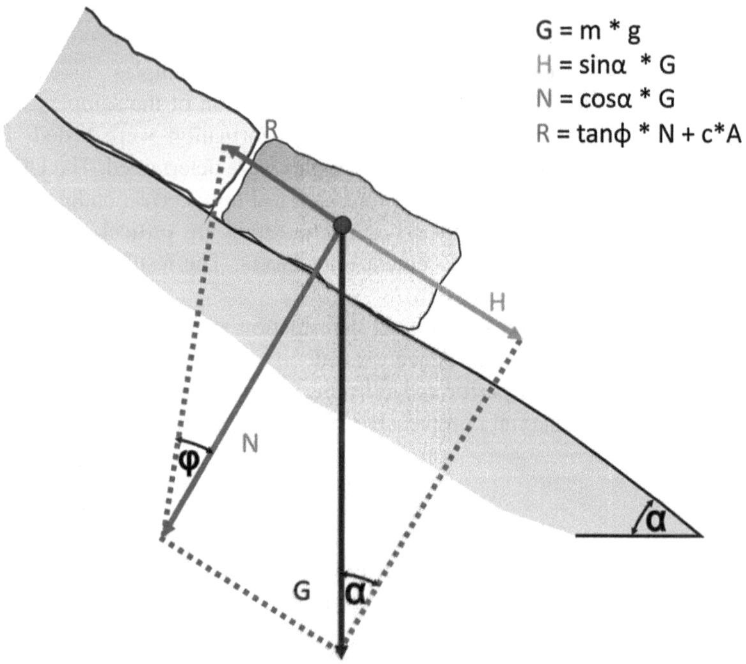

Fig. 4.5 Mechanical approach for a limit equilibrium analysis in case of planar failure. The formulas for the determination of the acting forces are given at the *right side*, where c is the cohesion and A is the total area underneath the block

Table 4.1 Scenario-matrix illustrating the approaches considered for the limit equilibrium analysis of the critical block at the Wachterl-Horn

	Minimum block volume (morphological visible block)	Maximum block volume (taking the dimensions of the failure surface into account)
Mechanical approach for a block on an inclined plane	Scenario 1	Scenario 2

The differences in width of the block dimensions result from field observations. The considered scenarios are labelled with numbers

named with G = force of gravity, N = Normal force, R = Retaining force and H = driving force.

As one of the basic parameters, the dimensions of the block were determined in the field, measuring the length of the block by abseiling using a measuring tape and a folding rule. The width was measured from two perspectives: by measuring the outer dimensions and the inner ones directly at the failure surface underneath the block, which results in two volume-scenarios for the limit equilibrium analysis (Sect. 5.1.5). For calculating the mass and further the force of gravity, we assumed a density of the dolomized limestone of 2830 kg/m^3, based on laboratory testing (Sect. 4.1.4). The basic friction angle was assumed as 35° according to the data of limestone-samples (Heckmann et al. 2012; Cruden and Hu 1988; Hoek et al. 1998).

Based on the information given above, we decided to consider the following scenario matrix taking the varying width for the limit equilibrium analysis into account (Table 4.1).

The mechanical approach was computed first without the consideration of cohesion. In a second step the necessary cohesion was back calculated for a labile stage with a Factor of Safety of 1.0, where the driving forces equal the retaining forces. The aim was to assess the content of cohesion necessary to stabilize the system at least to a labile stage of equilibrium.

The results of the stability analysis are described in detail in Sect. 5.1.5.

Further we analyzed the influence of the shear parameters (joint roughness coefficient and joint compressive strength) on the dilation angle "i" (Barton and Choubey 1977). The influence of the JRC on the dilation angle "i" was evaluated for three values of joint compressive strength obtained from JCS field-testing: Min = 20 MPa, Mean = 40 MPa and Max = 70 MPa. The failure criterion of Barton and Choubey (1977) represents the dependence of the shear strength on the normal stress, the fracture roughness and the cohesion:

$$\tau = \sigma_n \tan\left[\mathrm{JRC}\log_{10}\left(\frac{\mathrm{JCS}}{\sigma_n}\right) + \varphi_b\right] \quad (4.4)$$

$$i = \mathrm{JRC}\log_{10}\left(\frac{\mathrm{JCS}}{\sigma_n}\right) \quad (4.5)$$

where τ is the peak shear strength (assumed as the driving force normalized to the total contact area), σ_n is the effective normal stress, JRC is the joint roughness coefficient, JCS is the joint wall compressive strength and φ_b is the basic friction angle. The aim was to achieve a rough estimation which minimum JRC is necessary to compensate the offset of 20° between the assumed basic friction angle of 35° (Heckmann et al. 2012; Cruden and Hu 1988) and the slope angle of 55° at the release area.

4.1.6 Reconstruction of a Potential Mode of Failure

Based on the data obtained from field investigation a mechanical approach of a potential mode of failure was developed for the considered block. The mechanical interpretation is based on the assumption of a geometrical 2D-model suggesting the block on the inclined plane as the active block and the toe of the block as the passive block.

Thus the complex situation from field work was in a first approach reduced to a simple mechanical model of two key-blocks influencing the detachment process. For the generalized scenario of these two key-blocks a step-by-step model was developed to illustrate the progress of detachment (Sect. 5.1.6). The mechanical interpretation contains the observations from field work including the deformation process of the two blocks as well as the acting forces.

4.1.7 Monitoring of the Critical Block: Installation of Strain Gauges

To perform a continuous monitoring of the displacement rates, four mA-strain gauges were installed along the failure surface: One strain gauge at the outer failure surface and three strain gauges inside the failure cave, underneath the block (Fig. 4.6). The strain gauges have a surveying precision of hundredth of mm. The signal is recorded in mA and afterwards transformed in mm length. The strain gauges were installed almost equally distributed over the failure surface. The idea was to obtain knowledge about the behavior of deformation, meaning to assess whether the north or the south part of the block is moving to an increased extend. The sampling rate was set to one measurement per hour; the first test period was half a year from October 2012 to April 2013. This measuring period was the single period for deformation measuring due to a crash of the data logger system.

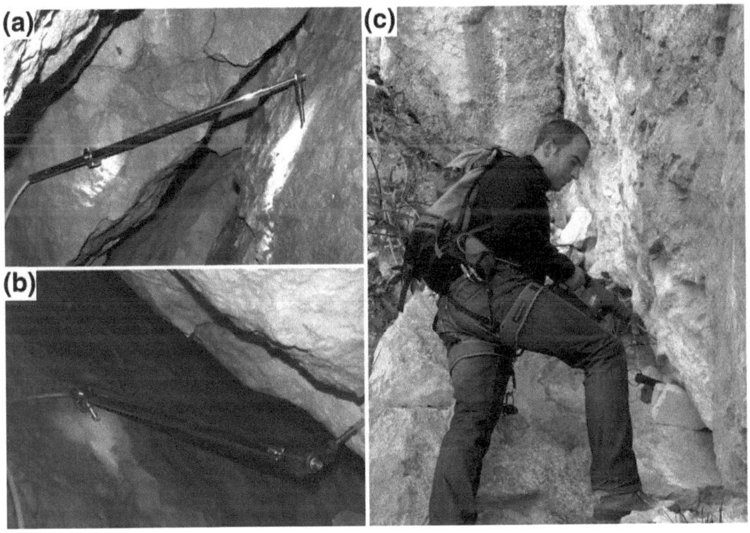

Fig. 4.6 Installation of the strain gauges: **a** strain gauge 3, installed in the middle of the failure cave; **b** strain gauge 4, installed at the end of the accessible area inside the failure cave; **c** Prof. Dr. Michael Krautblatter installing strain gauge 1 at the outside of the critical block

4.2 Quantitative Magnitude Assessment

To yield reconnaissance about a potential degree of fragmentation the information obtained from the source area investigation was combined with the recording of block dimensions at the talus slope. The first part of this chapter deals with the study of the source area, in the second part of the current chapter the quantitative recording of block sizes at the talus slope is described.

4.2.1 *Scanline Analysis*

The scanline analysis was divided in two sections according to the purpose of data analysis. For a statistical evaluation of the discontinuity orientations the recorded joints along the project site were evaluated and weighted referring to the scanline (Priest 1993). 105 joint orientations were taken into account to characterize the main discontinuity sets along a "virtual" scanline, determined in ArcGIS (Fig. 4.7). The scanline is called a "virtual" scanline since the recording distance along the entire project site would have been too long for direct recording along a measuring tape. Thus it was decided to develop a "virtual" scanline orientation along the line of recording across the slope. The "virtual" scanline orientation corresponds to the average orientation of the forest road crossing the slope. The aim of evaluating the

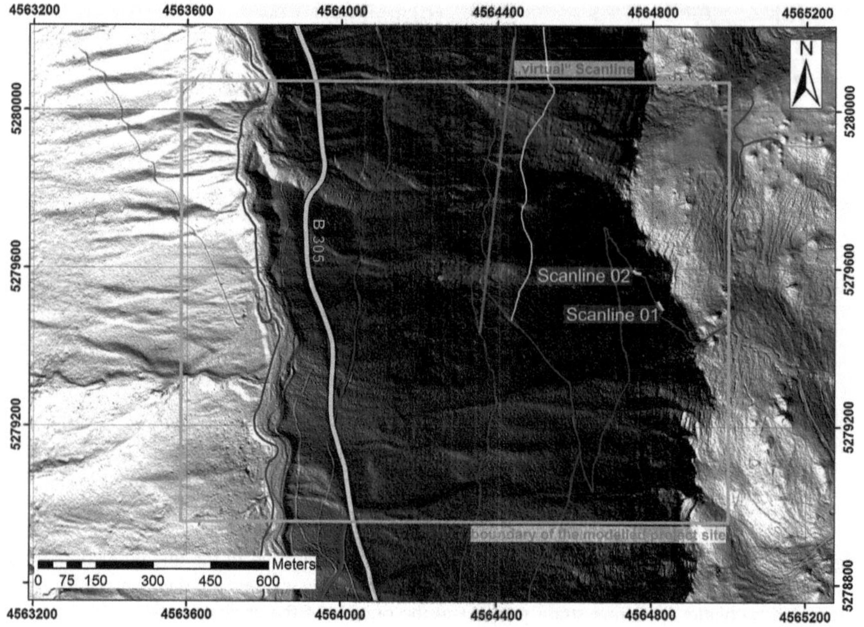

Fig. 4.7 Location of the "virtual" scanline along which the dominant discontinuity sets at the project site were recorded. Due to the outcrop conditions the scanline exceeds the edge of the modelled project site

discontinuities along this scanline was to create basic data for the following kinematic analysis in relation to the rock face of the source area.

For the more detailed discontinuity analysis around the critical block two 20 m scanlines were recorded at the vicinity of the block. Due to the limited accessibility and the outcrop conditions it was not possible to record more than the two scanlines at this part of the source area. For the detailed scanline sections the following parameters were recorded (Priest 1993) (see Table 4.2).

It is to mention that scanline 2 had to be divided in two subsections due to the variation in scanline orientation. In consequence for the presentation of the statistical joint analysis the results of scanline 1 were selected.

Due to the stepped terrain at the carbonate rock cliffs the lower half trace length was not recorded. The discontinuity spacing was determined along the scanline as

Table 4.2 The parameters recorded during the scanline analysis at the critical block, visualized as a header of the recording form

Dip	Dip direction	Joint distance	Upper half Trace length	Joint roughness coefficient	Curvature	Sort of discontinuity (bedding, joint, fault, cleavage)

Fig. 4.8 Scanline recording at the source area: **a** scanline 01 with a total length of 20 m, view from *north* to *south*; **b** the recording of the dominant joint systems, here as a showcase *marked with colors*; **c** recording of the joints orientation along the scanline

well as perpendicular to the joints. Vertical scanlines were not recorded at the study site due to the hard accessibility and partly high vegetation rates at the rock faces. Figure 4.8 shows the conditions of scanline recording at the source area.

4.2.2 Joint Systems and Kinematic Analysis

The kinematic analysis was performed using DIPS 6.0 (Rocscience) and incorporated the failure modes of planar sliding, wedge failure and direct toppling in relation to the rock face. For each failure analysis a minimum friction angle of 35° was assumed for the carbonates of the Dachstein-Formation (Cruden and Hu 1988; Heckmann et al. 2012; Hoek et al. 1998). The analysis of planar failure is based on the pole points of critical failure planes dipping in slope direction (Markland 1972 in John and Deutsch). The analysis of wedge failure is based on the orientation of the intersection linears dipping in slope direction, including the friction angle (Talobre 1957 in John and Deutsch). The direct toppling analysis refers to the intersections of two joint sets dipping into the slope, providing the degree of freedom for block toppling and one basal plane, providing a potential sliding plane.

The kinematic analysis of planar and wedge failure was in addition validated by hand using a Schmidt Net (equal area, lower hemisphere projection). The analysis was based on the evaluated data from scanline analysis (Sect. 5.2.1) to account for

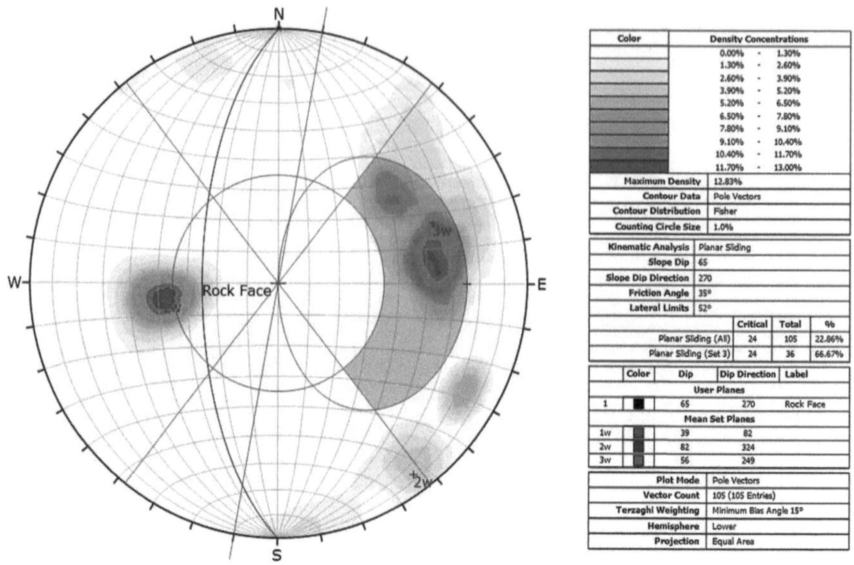

Fig. 4.9 Screenshot of the kinematic analysis tool of DIPS (Rocscience) for the example of a planar failure scenario. The critical area (Markland Plane) is coloured in *red*

the mean values of the discontinuity sets. The data analysis was performed for all three scanlines, where the results of the "virtual" scanline will be visualized in Sect. 5.2.2 due to the most joint recorded joints.

As an example of data evaluation on DIPS 6.0 (Rocscience) Fig. 4.9 shows a screenshot of the kinematic analysis tool, here for planar sliding. The black great circle represents the rock face; the associated Markland plane is marked with the red area. The joint data are plotted as a contour plot, indicating the density concentrations of the determined joint sets.

4.2.3 *Quantitative Evaluation of Block Dimensions at the Talus Slope*

In addition to the discontinuity analysis at the source area a quantitative block recording at the talus slope was performed. For each block the three block axes were recorded, where the x-axis is the longest, the y-axis is the middle and the z-axis corresponds to the shortest axis. The mean obstacle height (MOH) was measured for each block using a folding rule, which implies the height of an obstacle measured at the upslope side of the obstacle (Dorren 2012; Fig. 10).

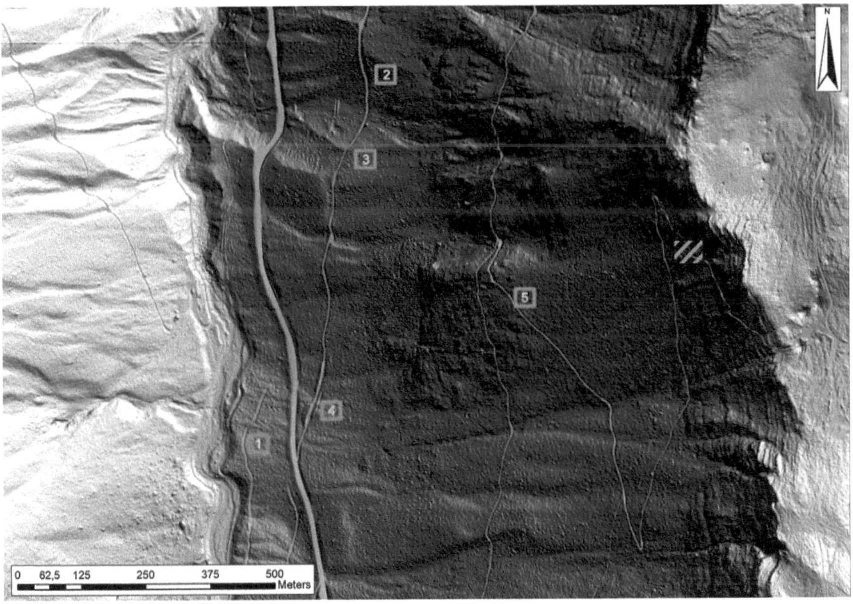

Fig. 4.10 The block dimensions were recorded quantitatively in 5 sample areas distributed over the proximal and distal talus slope. Each sample area measured 20 × 20 m; the three block axes (x, y and z) as well as the mean obstacle height were recorded

The dimension-recording was performed in sample areas of 20 × 20 m (Dorren 2012), where only the blocks with a mean diameter larger than 0.1 m were recorded. Rocks with a mean diameter smaller than 0.1 m are suggested to have once a minor influence on the surface roughness and second can be pressed into the ground during the block-slope-interaction, especially in case of vegetated slopes. In total we evaluated 5 sample areas which are consistently distributed across proximal and distal areas of the talus slope, to include the deposits from the valley bottom as well as the ones from the talus slope (Fig. 4.10).

Figure 4.11 shows some impressions from the block dimension recording at the talus slope; the block dimensions were determined using a folding rule and a measuring tape.

The results contain the evaluation of the mean diameter and the mean obstacle height (MOH) in Sect. 5.2.3, in order to compare both parameter sets for each sample area.

The results from the block-dimension recording provide basis parameter sets for the block axes and shapes used for the subsequent rockfall modelling using the codes Rockyfor3D and RAMMS::Rockfall.

Fig. 4.11 Evaluation of block dimensions at the talus slope. *Left* Overview of a sample area at the mid talus slope. (*Right, Top*) Typical compact block shape of the Dachstein carbonate material. *Bottom* slope cut along a forest road section, exposing the talus material at the surface

4.3 3D Rockfall Modelling

In the current Ph.D. thesis two 3D rockfall codes were selected for analyzing the rockfall hazard potential along the federal road B 305. Using the code Rockyfor3D the rockfall potential across the whole project site was considered, whereas the code RAMMS::Rockfall was chosen for modelling a mid-magnitude event of the critical block subjected to planar failure (Sect. 4.1). The following chapter will be divided into two subsections:

- The determination of the homogenous areas based on field investigation and geodata analysis
- The run-out modelling including the input parameter determination and the modelling itself using Rockyfor3D and RAMMS::Rockfall.

4.3.1 Determination of the Homogenous Areas Based on Field Investigation

The homogenous areas represent regions of invariant input-parameters for the current code. The homogenous areas were clustered referring to the ground parameters, the parameters of the source area and the forest stand parameters.

Table 4.3 General set-up of the parameters for the rockfall simulations related to the process areas

	Source area	Transit area	Run-out area
Slope roughness		×	×
Slope damping		×	×
Maximum block volumes	×		
Block shape and volume	×	×	×
Forest stand		×	×

In Table 4.3 the general parameter sets for 3D rockfall modelling are summed up in the column and related to the process area (rows) where the parameters are recorded.

The homogenous areas have been mapped combining the information from field work and geodata, including the 1 m-digital terrain model (DTM) and aero photos.

Source Area

The extension of the rock cliffs was determined, combining the information of hillshade maps generated of a 1 m-DTM and extensive field work (Fig. 4.12). A map of the slope inclination angle was created in ArcGIS to include the critical

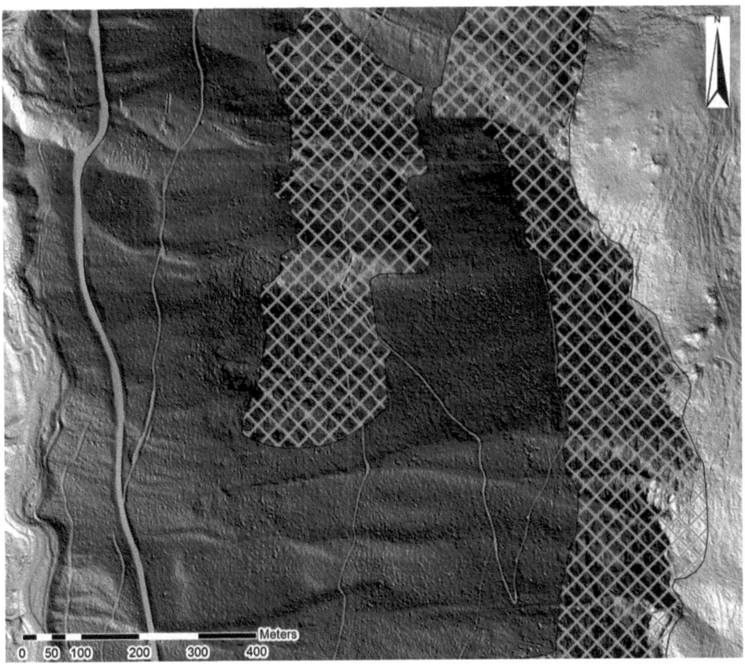

Fig. 4.12 Hillshade map of the project site with an illumination direction of 315°. The source areas are marked with the *red* cross-*dashed* fillings

inclination angle of 45° for the rockfall disposition map. The most essential parameters for the source area are the released block dimensions/axes and the block shape. The detailed decoding of these two parameter sets depends on the chosen modelling code. A detailed approach of block dimension assessment at the talus slope is described in Sect. 4.2.3.

Transit- and Deposit Area
The areas between and beneath the two crucial rock cliffs were defined as transit-or/and run-out areas.

The homogenous areas for the slope parameters were determined according to the following parameters:

- Ground parameters:
  - Slope Roughness parameters
    These depend on the amount and size of obstacles in general, for example fragmented rock material at the talus slope, tree stems or root stocks. In relation to the block size, the slope roughness affects the run out.
  - Damping referring to eight classes (Dorren 2012)
    The factor of damping affects the energy loss of the falling block due to the penetration depth into the slopes material.
- Forest stand (detailed parameters depend on the chosen code)

The slope parameters were assessed during field work using a fact sheet (Dorren 2012; Fig. 4.13). Every time the parameters of roughness or damping were changing a new homogenous area was set up. The forest stand parameters were determined using aero photos and information obtained from field work. The aero photos with a resolution of 0.2 m provide the possibility of counting the stock density as stems per hectare in ArcGIS. Afterwards the results have been validated in the field. At the Weißwand Wald significant areas of windblow occur, which lead to deficiency of the protection forest. The areas of windblow were assessed using aero photos combined with the information from field work.

4.3.2 Rocky for 3D

The aim of modeling rockfall scenarios during the project was to yield an enhanced assessment of block run-out along the federal road B 305 due to a considerable rockfall hazard in this road section.

For this purpose we performed detailed field work for the determination of the input parameters including a mapping of the process areas as well as the assignment of the homogenous areas. The following chapter is structured referring to the procedure of rockfall evaluation: field analysis and evaluation of Geodata—evaluation in GIS—modeling. The following description of parameter assessment and evaluation is based on Rockyfor3D (Dorren 2012).

Polygon characteristics								
1. Dominating rock (deposited in the polygon or potentially falling from release area)								
Block shape	☐ 1. rectangle		☐ 2. ellipsoid		☐ 3. Sphere		☐ 4. Disc	
Block dimensions (d1, d2, d3):	.... (m) x .... (m) x .... (m)							
Rock density (kg.m^{-3}):								
2. Soil / underground type in the polygon								
Material constituting the under-ground	☐ river / swamp / other mate-rial in which a rock could penetrate completely	☐ fine soil material (depth > ~100 cm)	☐ fine soil material (depth < ~100 cm) / sand/gravel mix in the valley	☐ scree (Ø < ~10 cm) / medium compact soil with small rock fragments / forest road	☐ talus slope (Ø > ~10 cm) / compact soil with large rock fragments	☐ bedrock with thin weathered material or soil cover	☐ bedrock	☐ asphalt road
(soiltype) values needed for Rockyfor3D	0	1	2	3	4	5	6	7

3. Surface roughness in the polygon		
MOH: typical obstacle height normal to the slope surface (m) that block encounters in 70%, 20% and 10% of the cases during a rebound on the slope surface. **Should be measured looking down the slope!** 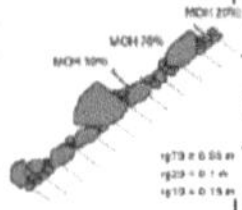	MOH for 70% of the sample area (rg70)	0 - 100 (m)
	MOH for 20% of the sample area (rg20)	0 - 100 (m)
	MOH for 10% of the sample area (rg10)	0 - 100 (m)
Lying tree stems*	Mean height = m	Area covered = %

4. Forest*			
Representative plot size: ...... m × ...... m			
DBH# (cm)	# DBH: Tree diameter at breast height Record all the DBH ≥ 5 cm measured in the plot: e.g., 8, 31, 17, 13, ...		
Stems / ha			
Mean DBH (cm)		Coniferous (%)	
Stddev DBH (cm)			
Species*			

Fig. 4.13 Fact sheet for determining the required input parameters for the code Rockyfor3D (Dorren 2012)

Field Investigation

During field work the parameters for each process area were determined according to a fact sheet (Dorren 2012). Relating the input parameters to the sections of process areas where they belong to, the following structure can be provided (Table 4.4).

Table 4.4 Input parameters for the code Rockyfor3D related to the process areas

Source area	Transit- and deposit area	Topography (whole project site)
Block axes (d1, d2, d3)	Roughness as Mean obstacle height (MOH) as rg 70, rg 20 ad rg 10	DEM with a resolution between 1.0 and 10 m
Block shape as 4 classes	Soiltype as 8 classes of damping categories (7 = asphalt, 0 = river swamp)	
Rock density	Forest stand as stock density, stock type, percent of coniferous trees, the diameter at breast height (DBH)	

Source Area
As described in Sect. 4.2 we evaluated the block dimensions and shape at the source area as well as at the talus slope. The block axes are represented by the parameters d1, d2 and d3 where d1 is defined as the longest axis and d3 as the shortest axis. The block shape is characterized by four shape categories: 1 = rectangular block, 2 = ellipsoidal block, 3 = spherical block and 4 = disc shaped block (Dorren 2012). The rockdensity is either based on literature values or on laboratory testing (Sect. 4.1.4).

Transit- and Deposit Area
The main two parameters characterizing the slope conditions are the roughness and the damping. The roughness is quantified by the mean obstacle height (MOH) representing the obstacle height measured at the upslope side of a certain block. Since a deviation of mean obstacle height should is included for each homogenous area, the percentiles of 70, 20 and 10 % of the MOH in one homogenous area is required. The damping is quantified according to 8 categories from 0 = river swamp to 7 = asphalt, which can be considered as an estimation assistance and are linked to the normal coefficient of restitution (R_n). In the current project the slopes roughness parameters were assessed during mapping and in terms of quantitative block recording (Sect. 4.2.3), the damping parameters were assessed on the basis of the mentioned classification.

Forest Stand
The forest stand is explicitly classified in Rockyfor3D. In the current project the stock density as stems per hectare was determined by counting tree crowns in aero photos with a resolution of 0.2 m and validating the information in the field. The percentage of coniferous trees and the stock type was roughly assessed by counting trees in sample areas of 20 by 20 m in the field. The diameter at breast height (DBH) was randomly measured in the field.

Transfer of the field parameters to a GIS Environment
The first step in using the code Rockyfor3D is setting up a GIS project, in which the recorded parameters from field work are visualized as map layers. The mapped areas represent the homogenous process areas (feature classes): source- and transit or deposit area and forest stand parameters. Each feature class contains the characterizing parameters related to the process area as fields, which are created as polygons in ArcGIS. The data structure for the current project is visualized in Fig. 4.14. In total at least 10 parameter raster layers are exported as ESRI ASCII files: the DEM, the block shape, the three block axes (each as an extra ASCII), the block shape, the rock density, the three roughness parameters (each as an extra ASCII) and the soiltype (damping).

In case of modelling with forest stand 4 additional ASCIIs: the stems per hectare, the DBH mean and standard deviation and the percentage of coniferous trees are required. All ASCII layers must all show the same area extent and resolution. If all ASCII files are provided in the modelling folder, the datasets can be loaded into the Rockyfor3D set up.

Fig. 4.14 Data structure of the file-geodatabase set-up for the modelling using Rockyfor3D. The feature classes source area, transit area and forest stand (*left*) contain the mentioned fields as parameter classes (*right*)

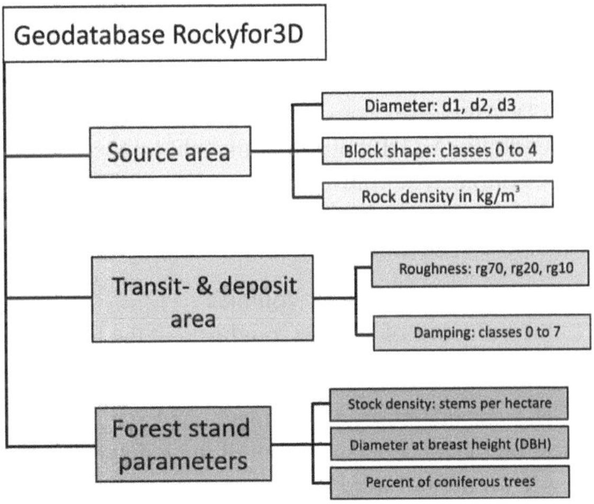

Modelling

The project site at the "Weißwand Wald" is described in detail in Chap. 3.

The Rockyfor3D model was set up, following a scenario matrix (Fig. 4.15). The matrix is divided into16 fields, combining different magnitude classes with four slope classes (varying in terms of roughness). The magnitude classes follow volume classes recorded quantitatively at the talus slope (Sect. 5.2.3): min = 0.15 × 0.2 × 0. 25 m; mean = 0.2 × 0.25 × 0.35 m; max. = 0.5 × 0.6 × 0.8 m. A fourth class of "blocks" (0.8 × 1.0 × 1.2 m) was integrated to include relatively large blocks following the hazard indication map (Bayerisches Landesamt für Umwelt 2014:

underground parameters / block volume	Min. 0.15x 0.2x 0.25 m	Mean 0.2x 0.25x 0.35 m	Max. 0.5x 0.6x 0.8 m	blocks 0.8x 1.0x 1.2 m
Transit_1 parameters based on fieldwork; validated ref. to Dorren (2010)				
Transit_2 increasing roughness (MOH) + 0.1 m at the talus slope + 0.15 m at the valley bottom				
Transit_rough_1 parameters based on fieldwork; validated ref. to Dorren (2010) incorporating areas of windbreakage				
Transit_rough_2 increasing roughness (MOH) + 0.1 m at the talus slope + 0.15 m at the valley bottom incorporating areas of windbreakage				

Fig. 4.15 Matrix set up for the rockfall simulations performed at the project site Weißwand using the code Rockyfor3D. The scenarios were chosen in terms of varying block volumes and roughness parameters including areas of windblow

14), but taking the rock shape into account. The slope-parameter-classes include four classes: Transit 1 represents the ground parameters recorded in the field by mapping, validated with the suggested ground parameter set provided by Dorren (2012). All classes named with transit_2 represent an increased roughness of 0.1 m added to the MOH mapped at the talus slope and 0.15 m added to the MOH at the alley bottom. The transit areas marked with the term "rough" additionally consider the increased roughness in terms of rootstocks and root plates in the areas of windblow. The aim the current modelling matrix was to consider:

1. the influence of block volumes
2. the influence of varying roughness at the talus slope and areas of windblow (root plates and broken tree stems)

The consideration of the mentioned scenarios is suggested to have an evident effect on the hazard assessment of the federal road B 305 at the bottom of the slope.

All scenarios were modelled with and without forest stand. The results of selected scenarios are presented in Sect. 5.3.1. The scenarios were chosen depending on the frequency of rockfall based on the recording of rockfall deposits and the hazard exposure of the federal road B 305.

4.3.3 RAMMS::Rockfall

The code RAMMS::Rockfall was used for the analysis of Mid-Magnitude Events and their run-out analysis. Due to large block volumes at the release locations these events incorporate an evident degree of fragmentation. To provide an approach, which links the degree of fragmentation to the rockfall run-out, we performed parameter studies taking different volume classes into account. The volume classes are based on the block cubatures recorded in the field: min. degree of fragmentation refers to the block volume at the source area and the max. degree of fragmentation refers to the block volumes at the talus slope.

Parameter Entry

In contrast to the code Rocky for 3D, RAMMS::Rockfall is not strictly based on a GIS approach, so the input parameters are directly entered in the Code interface, not imported form a GIS environment. The next paragraph is based on the following literature: Bartelt et al. 2013; Leine et al. 2013.

The input parameters are divided into two main sets: the descending material (blocks) and the slope material (see Table 4.5).

The topography is given by a Digital Elevation Model in ASCII Format.

The block shape is characterized by three main block shapes selectable in the rock builder (Fig. 4.16). Each rock shape is divided into three subtypes depending on the length to width ratio, where the length to width ratio increases with increasing label number.

Table 4.5 Input parameters for RAMMS::Rockfall divided by the parameter set they belong to

Parameter set	Parameter	Description of the parameter	Unit
Block	Volume	Determined from the three block axes	m^3
Block	Density	The density of the descending material	Kg/m^3
Block	Shape	Chosen out of three shape types: cuboid, flat or elongated; the proportion of the axes to each other can be entered via subtypes for each class	–
Block	κ (kappa)	As a coefficient for the determination of the blocks "Eigenframe" (the blocks coordinate system)	–
Slope	μ (Mu)	As a min. and max. value of the friction coefficient	–
Slope	β (beta)	The parameter controls how quickly the friction is released as the rock departs the slope. β is linked to the penetration depth of the rock.	–
Slope	ε (epsilon)	Corresponds to the coefficient of restitution in normal direction	–

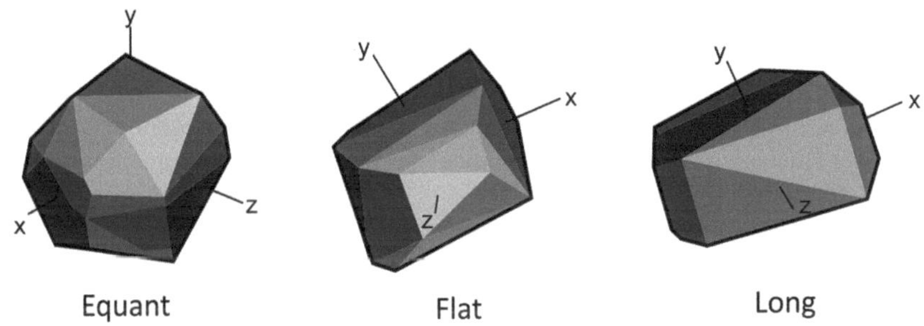

Fig. 4.16 The selectable *block shapes* in RAMMS::Rockfall. "Equant" is set to be the most compact one and "long" the most stretched block shape

The ground parameters are defined according to their special extension via shape files. The friction and ground drag parameters are defined directly in RAMMS:: Rockfall according to the slope-classes "very soft" to "extra hard".

In addition to the mentioned input parameters the forest stand can also be taken into account. The forest stand is characterized by the tree stem height and the tree drag. These two parameters vary depending on the chosen forest density. The parameter input follows the shapefiles for the special extension and is performed via parameter selection directly in RAMMS::Rockfall.

Since the current Ph.D. thesis deals with the applied aspects of rockfall modelling it was decided to take over the predefined underground parameter classes from RAMMS::Rockfall. Otherwise variables for the blocks coordinate system and additional angles would have been necessary to define.

Modelling

The aim of using RAMMS::Rockfall during the current Ph.D. project was the consideration of potential fragmentation scenarios of the critical block at the Wachterl-Horn by performing parameter studies. The details about the modelled block and the results from field investigation are described in Sect. 5.1.1. The information on rock density was determined in terms of immersion weighing during laboratory testing (Sect. 4.1.4).

The assumed scenarios are based on the recording of joint persistence in the field. It was decided to classify three main scenarios (min., mean and max. fragmentation), where the "maximum fragmentation" scenario is based on the block volumes assessed by the recording of joint persistence. The class "mean fragmentation" mirrors the volumes by halved block axes, and the class "minimum fragmentation" by quartered block axes of the class "maximum fragmentation". The class "minimum rockfall" was added to yield a contrast to the smallest possible rock volume, since 0.1 m^3 is the smallest volume being edited in RAMMS::Rockfall. The selected block shapes for each created block area visualized in Table 4.6. With decreasing block volumes we assumed a more compact ("equant") rock shape in the carbonates due to fragmentation and weathering processes. Only for scenario "minimum rockfall" we mixed up all selectable block shapes to see the shape effect in the volumes class.

The slope parameters were set referring to the ground parameter classes editable in RAMMS::Rockfall. For the Source area a "hard" terrain was selected; the talus slope was characterized by "medium hard" terrain and the accumulation area by "medium" terrain.

Based on the results of joint persistence eight different blocks with little changes in block shape were defined. The following block volumes with the associated block axes and shapes were defined.

Table 4.6 Visualizes the 4 determined block classes with the associated block shapes and volumes

Minimum rockfall		Maximum fragmentation		Mean fragmentation		Minimum fragmentation	
Shape	Volume (m^3)	Shape	Volume (m^3)	Shape	Volume (m^3)	Shape	Volume (m^3)
Equant_1.2	0.1	Equant_1.2	0.4	Equant_1.3	2.62	Flat_1.2	12.04
Equant_1.3	0.1	Equant_1.2	0.5	Equant_1.3	2.7	Flat_1.2	16.6
Equant_1.3	0.11	Equant_1.3	0.2	Equant_1.3	2.09	Flat_1.2	21
Flat_1.2	0.1	Equant_1.3	0.4	Flat_1.2	4.18	Flat_1.5	22.7
Flat_1.5	0.1	Equant_1.3	0.6	Flat_1.2	1.56	Flat_2.0	31.6
Flat_2.0	1.0	Flat_1.2	0.5	Flat_1.2	2.2	Long_1.2	33.1
Long_1.2	0.1	Flat_1.2	0.6	Long_1.2	3.94	Long_1.2	21.7
Long_1.2	0.11	Flat_1.2	0.7	Long_1.2	2.86	Long_1.2	17.6

References

Bartelt P, BuehlerY, Christen M, Dreier L, Gerber W, Glover J, Schneider M, Glocker Ch, Leine R, Schreizer A (2013) RAMMS, User Manual v1.5 Rockfall. p 83 (pdf-version)

Barton N, Choubey V (1977) The shear strength of rock joints in theory and practice. Rock Mech 10: 1–54

Bayerisches Landesamt für Umwelt (LfU) [Hrsg.] (2014) Gefahrenhinweiskarte Alpen mit Alpenvorland—Landkreis Berchtesgadener Land. 78S., München (Bayrisches L.-Amt für Umwelt, pdf-version)

Cruden DM, Hu XQ (1988) Basic friction angles of carbonate rocks from Kananaskis country, Canada. Bull Int Assoc Eng Geol 38:55–59

DGGT—Deutsche Gesellschaft für Geotechnik e.V (2004) Einaxiale Druckversuche an zylindrischen Gesteinsprüfkörpern. Neufassung der Empfehlung Nr. 1 des Arbeitskreises Versuchstechnk Fels der Deutschen Gesellschaft für Geotechnik e.V.- Bautechnik 81:825–834

DIN 52102 (2012) Prüfverfahren für Gesteinskörnungen—Bestimmung der Trockenrohdichte mit dem Messzylinderverfahren und Berechnung des Dichtigkeitsgrades. 10 S., Berlin (Beuth-Verlag)

DIN 51220 (2003) Werkstoffprüfmaschinen- Allgemeines zu Anforderungen an Werkstoffprüfmaschinen und zu deren Prüfung und Kalibrierung, Berlin (Beuth Verlag)

DIN EN 1097-6 (2013) Prüfverfahren für mechanische und physikalische Eigenschaften von Gesteinskörnungen—Teil 6: Bestimmung der Rohdichte und der Wasseraufnahme; Deutsche Fassung EN 1097-6:2013, Berlin (Beuth-Verlag)

Dorren LKA (2012) Rocky for (v5.0) revealed—Transparent description of the complete 3D rockfall model. ecorisQ paper (www.ecorisq.org) 31 p

Heckmann T, Bimöse M, Krautblatter M, Haas F, Becht M, Morche D (2012) From geotechnical analysis to quantification and modelling using LiDAR data: a study on rockfall in the reintal catchment, Bavarian Alps Germany. Earth Surf Process Landforms 37:119–133

Hoek E, Marinos P, Benissi M (1998) Applicability of the geological strength index (GSI) for very weak and sheard rock masses. The case of the Athens Schist Formation. Bull Eng Geol Env 57:151–160

ISRM—International Society for Rock Mechanics (1978) Suggested methods for the quantitative description of discontinuities in rock masses. Commission on standardization of laboratory and field tests. Int J Rock Mech Min Sci Geomech Abstr 15 4:319–368

John KW, Deutsch R (1974) Die Anwendung der Lagenkugel in der Geotechnik.-43 S., unveröff. Skriptum (online-pdf), Institut für Felsmechanik, Technische Universität Graz

Leine RI, Schweizer A, Christen M, Glover J, Bartelt P, Gerber W (2013) Simulation of rockfall trajectories with consideration of rock shape. Multibody Syst Dyn, p 31, Retrieved as pdf-version at Springer Science + Business Media, Dordrecht, doi: 10.1007/s11044-013-9393-4

Markland JT (1972) A useful technique for estimating the stability of rockslopes when the rigid wedge sliding type of failure is expected. Imperial College, Rock Mech. Research Report, No. 19: 10p

McCarroll (1997) A template for calculating rock surface roughness. Earth Surf Process Landforms 22:1229–1230

McCarroll D, Nesje A (1996) Rock surface roughness as an indicator of degree of rock surface weathering. Earth Surf Proc Land 21:963–977

Priest SD (1993) Discontinuity analysis for rock engineering. Chapman & Hall, London, 473p

Talobre J (1957) La méchanique des roches. Dunop, Paris S39–S44

Thuro K (1996) Bohrbarkeit beim konventionellen Sprengvortrieb. Münchner Geol Hefte Reihe B: Angewandte Geologie: Heft 1 (elektronische Publikation) 145S

Thuro K, Plinninger RJ, Zäh S, Schütz S (2001) Scale effects in rock strength properties. Part 1: Unconfined compressive test and Brazilian test.- ISRM Regional Symposium Eurock, June 3–7, Espoo, Finnland (ISBN 90 2651 821 8, pdf-version)

Tse R, Cruden DM (1979) Estimating joint roughness coefficients. Int J Rock Mech Min Sci Geomech Abstr 16:303–307

Woszidlo H (1989) Untersuchungen an Festgesteinen mit dem Prallhammer nach Schmidt. Nat Tag Ing-Geol Bensheim Ber 7:287–294

Yang ZY, Lo SC, Di CC (2001) Reassessing the joint roughness coefficient (JRC) Estimation using (Z2). Rock Mech Rock Eng 34(3):243–251

Chapter 5
Results

The results section will be subdivided analogue to the methods section. Section 5.1 deals with the case study referring to the potential planar rock failure at the Wachterl-Horn. Section 5.2 illustrates the results of discontinuity analysis and the quantitative magnitude assessment and the third part will refer to the implementation of the magnitude information into 3D-rockfall simulations (Sect. 5.3).

For Sect. 5.1.1 "Recording of the block dimensions and assessment of joint persistence" the results of field recording are connected to field data interpretation to provide an integral data presentation. For the rest of chapters the interpretation will be presented in Chap. 6.

5.1 Case Study: Potential Planar Rock-Slide at the Wachterl-Horn

5.1.1 Field Investigation of the Critical Block and Its Vicinity

In the following section the results of the parameter recording at the mid-magnitude block subjected to planar failure will be presented. The section will start with the mapping of the block dimensions and the joint persistence. Due to the unique occasion of an accessible failure surface the results of the mapping of the failure surface and the recording of shear parameters will be described in the subsequent section.

Recording of the Block Dimensions and Assessment of the Joint Persistence
The detailed recording of joint persistence suggests that in case of failure the block would be divided at least into 8 single blocks (Fig. 5.1). As the kinematic analysis of Sect. 5.2.2 shows, the bedding is dipping towards the slope, one joint set is dipping parallel to the slope and the third joint set is oriented approximately perpendicular to the other both. The block dimensions can be specified with 9.0 m in length, 4.7 m in width across the slope and 3.3 m in height. Upslope of the block three fragmented blocks of a total length of 2.70 m (in slope direction) are located.

B. Sellmeier, *Quantitative Parameterization and 3D-run-out Modelling of Rockfalls at Steep Limestone Cliffs in the Bavarian Alps*, Springer Theses,
DOI 10.1007/978-3-319-24510-2_5

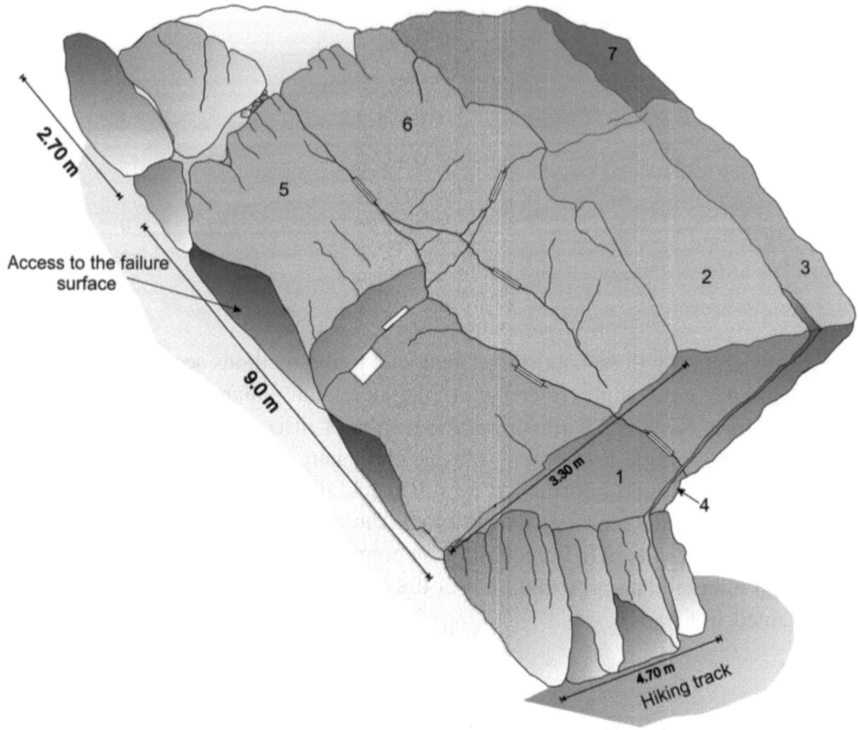

Fig. 5.1 Illustration of a suggested fragmentation scenario of the critical block at the Wachterl-Horn. The sketch suggests that the block will be at least divided into 8 single blocks due to the joint orientation and persistence

The blocks are loosened and have no visible rock-rock connection to the rock mass, but nevertheless would affect the driving forces. The "Müller Flags" illustrate the joint orientations in relation to the rock surface. The toe of the block, symbolized by the grey rock wedge below block 1 and 4 is suggested to act as a support for the whole system. The mechanical situation of a potential failure scenario is described in detail in Sect. 5.1.5.

Investigation of the Failure Surface

Changing the perspective and having a look underneath the block, the map of shows a plan view of the detachment surface (Fig. 5.2). The entrance to the failure surface at the northern rock face of the block is defined as the origin of the coordinate system. The up-slope limit of the cave is defined by cohesive or frictional contacts (rockbridges), representing the contact between the block and the rock mass. The base of the cave is formed by loosened dolomized limestone blocks ("blocky rock mass" in the sketch). The cave dimensions are approximately 10.5 m across slope and up to 6 m up slope. The rear 2 m of the cave is not accessible, since the height as well as the opening width of the cave decrease. The shaded polygons

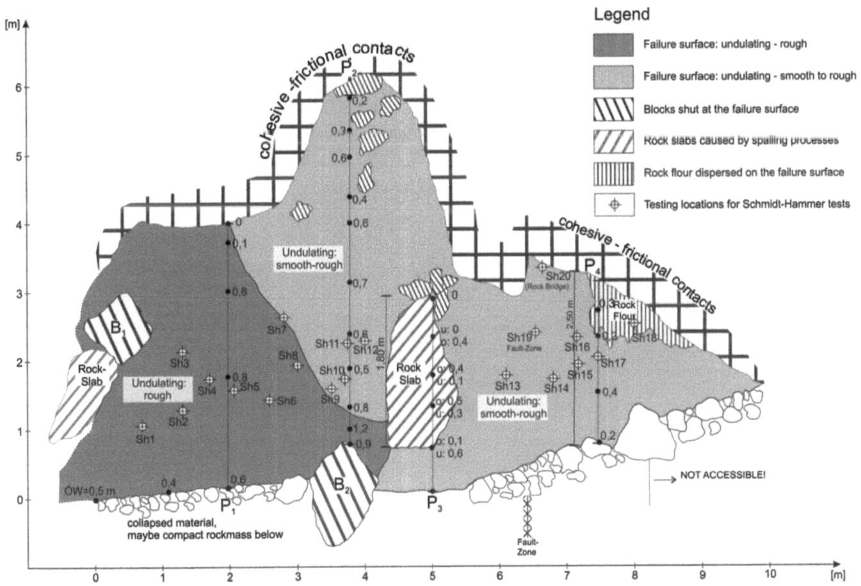

Fig. 5.2 Map of the failure surface of the observed block. The *color* indicates the roughness referring to ISRM (1978), the *black lines* represent cross sections through the cave labeled with the opening width in meter. The *vertically black* striped field on the right marks an area where sheared material (rock flour) was sampled

represent limestone blocks or slabs locked at the failure surface between block and rock mass.

The detachment surface can be divided in two sections based on fracture roughness (Fig. 5.2). The north part (left part in Fig. 5.2) can be characterized by its undulating rough fractures. The south part (right part in Fig. 5.2) can be described as undulating (ISRM 1978), but smoother than the left section. The vertically black-striped area in the right part of the cave marks a location where fine grained light grey sheared material is dispersed on the detachment surface, providing evidence that an active cracking of rock-rock contacts related to shearing failure takes place underneath the block. The cross-wire symbols mark the locations where Schmidt-Hammer testing was performed (Sect. 4.1.2)

The black lines (P_{1-4}) represent recorded cross sections through the cave labeled with the opening width. The recorded cross-sections are shown in Fig. 5.3. Our measurements along the profiles indicate that the opening widths vary between 0.1 and 0.5 m in the upslope-part and 0.5 and 1.2 m in the base-part of the cave. The blocks and nappes marked in the cross sections show that spalling processes take place underneath the block, leading to a variation in surface roughness at the block as well as at the rock mass surface. Nevertheless the illustration of the cross-sections suggests an evident coincidence of the shape of the block and the rock mass.

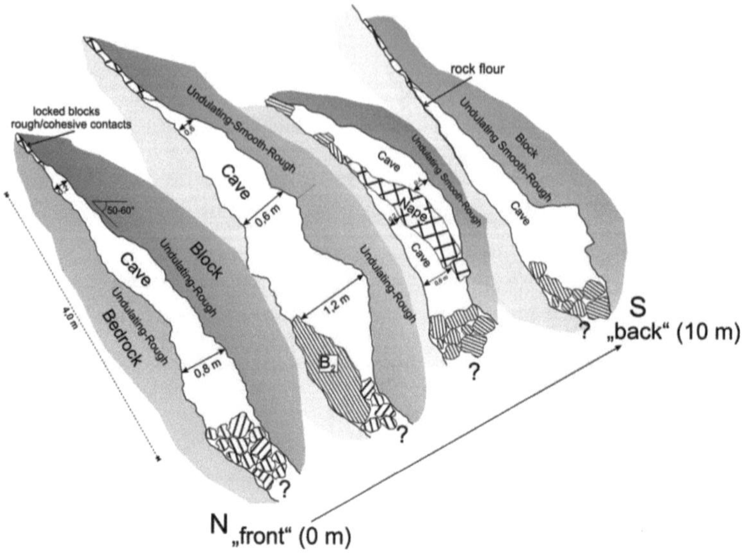

Fig. 5.3 Back-to-back-illustration of the cross sections along the failure surface. The sketch demonstrates the opening widths in dip direction of the failure cave. The nappes and block suggest that active spalling processes occur underneath the block

5.1.2 Recording of the Shear Parameters at the Failure Surface

As basic parameters for the subsequent limit equilibrium analysis (Sect. 5.1.5) we performed a quantitative determination of the joint wall compressive strength (JCS) as well as of the joint roughness coefficient (JRC).

The JCS tests were accomplished in situ at the detachment surface of the observed block. The test-positions are marked as Sh_{1-20}, where the distance from the cave entrance increases with increasing labeling of the test locations (Fig. 5.2). The results are evaluated in two ways: the lower/upper quartile and the median as the data basis for the Box-Whisker Plot; and the arithmetic mean values plotted versus the distance from the cave entrance. The results for the quartiles and the median values are presented in Table 5.1. For the evaluation of this Box-Whisker-Plot, we included all ten single values of each test series, meaning we were not performing any data correction (Barton and Choubey 1977). Since we intend to consider the detachment surface of the rock mass in in situ, we decided to take all single values of every test series into account. The mean values are visualized in Table 5.2, which are presented in two ways: the arithmetic mean value of the 10 single tests of one series and the results corrected referring to Barton and Choubey (1977). This correction requires the 5 lowermost values to be eliminated, implying that only the best 5 out of 10 values are plotted. The related plots are presented in Sect. 6.1.2.

Table 5.1 Shows the composition of the Schmidt Hammer data, used for the Box-Whisker Plot (Sect. 6.1.2)

In (MPa)	SH 1	SH 2	SH 3	SH 4	SH 5	SH 6	SH 7	SH 8	SH 9	SH 10
Lower quartile	30.5	23.5	26	30.5	24.5	27	27	28.5	47	37.5
Median	34	25.5	36.5	40	32.5	32.5	32.5	42	64.25	44.5
Upper quartile	41	27.5	80	46	47.5	50	41	45	85	66.5
	SH 11	**SH 12**	**SH 13**	**SH 14**	**SH 15**	**SH 16**	**SH 17**	**SH 18**	**SH 19**	**SH 20**
Lower quartile	30.5	23.5	26	30.5	24.5	27	27	28.5	47	37.5
Median	34	25.5	36.5	40	32.5	32.5	32.5	42	64.25	44.5
Upper quartile	41	27.5	80	46	47.5	50	41	45	85	66.5

The table shows the lower quartile, the median and the upper quartile

Table 5.2 Shows the composition of the arithmetic mean values of the Schmidt-Hammer tests

In (MPa)	SH 1	SH 2	SH 3	SH 4	SH 5	SH 6	SH 7	SH 8	SH 9	SH 10
Mean	35.5	25.8	45.8	38.85	35.5	38.5	34.7	37.9	67.6	53.1
Mean Barton and Choubey (1977)	40.6	27.3	56	46.8	46.8	50.4	43.5	44.8	93.2	69.3
	SH 11	**SH 12**	**SH 13**	**SH 14**	**SH 15**	**SH 16**	**SH 17**	**SH 18**	**SH 19**	**SH 20**
Mean	52.75	36.5	33.35	30.16	64.5	46.45	67.00	25.65	35.65	20.5
Mean Barton and Choubey (1977)	71.8	46.7	38.6	36.21	87.5	57.4	84.5	26.7	39	20.1

The *upper line* represents the mean values out of 10 single tests per series, the *bottom row* mirrors the mean values corrected referring to Barton and Choubey (1977), meaning the mean of the best 5 values

The JRC recording was performed along the cross sections P1, 2 and 4 at the failure surface, excluding P3 since this cross section includes a rock slab which was fallen into the failure cave due to spalling processes. The joint roughness coefficient was evaluated according to three intervals of measurement for each cross section. Each of the three cross sections consists of 10 to 12 clustered subsections (belonging to the length of the Barton-Comb). The values of Table 5.3 represent the average mean values of the JRC for each cross section for the measuring intervals of 5, 10 and 20 mm. In our roughness evaluation we did not assess the undulation across the failure surface in a quantitative way. Classifying the undulation visually, the northern area (left part) of the failure surface shows a much higher undulation than the southern area (right part) (ISRM 1978).

Table 5.3 Evaluation of the JRC measurements along three cross sections in dip direction of the failure surface

	Cross section P1	Cross section P2	Cross section P4
JRC for D(x) = 5 mm	18	17	16
JRC for D(x) = 10 mm	15	14	14
JRC for D(x) = 20 mm	13	11	10

The specification of D(x) represents the length of the measuring interval. At cross section 03 it was not possible to take roughness values because of a rock slab

Table 5.4 Visualisation of the results of the 10 uniaxial compressive strength tests

Sample-Nr.	Uniaxial compressive strength		Strain modulus	Destruction work
	σ_u (MPa)	Classification referring to ISRM 1978	V (GPa)	W_z (kJ/m^3)
1	84.3	High	42.61	191.4
2	105.2	Very high	45.16	201.8
3	81.9	High	40.84	116.4
4	85.6	High	42.54	140.9
5	108.7	Very high	44.26	187.6
6	93.1	High	46.95	118.3
7	88	High	44.14	79.8
8	91.4	High	41.67	136.1
9	115.3	Very high	–	212.6
10	92.6	High	38.92	159.7
Mean value	94.61	High	43.01	154.46
Minimum	81.9	High	38.92	79.8
Maximum	115.3	Very high	46.95	212.6

DGGT (2004), ISRM (1978), modified by Thuro (1996)

5.1.3 Uniaxial Compressive Strength Tests

The uniaxial compressive strength tests show a variation in strength from a minimum of 81.9 MPa to a maximum of 115.3 MPa (Table 5.4), which corresponds to values in the classes "high" and "very high" referring to ISRM (1978), modified by Thuro (1996). Three of 10 samples belong to the class "very high" and seven samples belong to the class "high". The strain modulus was determined at the linear elastic pre-failure curve sections. The results of the strain modulus vary between a minimum of 38.92 GPa to a maximum of 46.95 GPa. The destruction work represents the integral below the stress-strain-curve, which is in the current case the pre-failure curve including the point of failure. The results of the destruction work vary between minimum of 79.8 kJ/m^3 and a maximum of 212.6 kJ/m^3.

5.1.4 Determination of Density

The evaluation of the density determination under upwelling conditions showed the results illustrated in Table 5.5 for the 5 samples of the Dachstein-Formation.

For the subsequent stability analysis and the 3D rockfall modelling we assumed a density mean material density of 2.83 g/cm^3.

Table 5.5 Evaluation of the density parameters for the 5 carbonate samples of the Dachstein-Formation

	Min.	Mean	Max.
Dry bulk density (Trockenrohdichte) (g/cm^3)	2.69	2.83	2.97
Density (g/cm^3)	2.69	2.82	2.96

5.1.5 Limit Equilibrium Analysis

In the following section the results from the stability analysis of the critical block at the Wachterl-Horn are presented. The section will be divided into the scenarios 1 to 4 according to the scenario matrix of Table 4.1.

Scenario 1 represents the minimum assumed block volume with a block width of 4.50 m (across the slope) and was analyzed by the mechanical approach of a block on an inclined plane. The results of the mechanical limit equilibrium analysis without taking a cohesion part into account are provided in Table 5.6. The driving forces can be named as 3766 kN whereas the retaining forces are 1846 kN. The Factor of Safety for this case can be specified as 0.5; instable. For a labile stage of the system we back-calculated the cohesion part assuming a factor of safety of 1, where the retaining force equals the driving force. The last line of Table 5.6 shows the minimum cohesion, necessary for a labile stage, which can be specified as approximately 37 kN/m^2.

Scenario 2 takes the increased block volume considering a width of 6.5 m, based on the mapping of the failure surface, into account (Sect. 5.1.2). The width is

Table 5.6 Results of the limit equilibrium analysis for case 01, the smallest morphological block analyzed with the mechanical approach for a block on an inclined plane

Description	Formula	Unit	Value
Length	–	(m)	11.5
Width	–	(m)	4.5
Height	–	(m)	3.2
Contact area	A = L*B	(m^2)	51.75
Density	–	(kg/m^3)	2830
Slope angle	–	(°)	55
Assumed friction angle	–	(°)	35
Volume	V = l*w*h	(m^3)	165.6
Mass	M = ρ*V	(kg)	468,648
Force of gravity	G = m*g	(N)	4,597,436
Normal force	N = cosα*G	(N)	2,636,981
Driving force	D = sinα*G	(N)	3,765,999
Retaining force	R = tanφ*N	(N)	1,846,434
Factor of safety	**υ = R/D**	–	**0.5**
Minimum of cohesion for a labile stage (FS = 1)	c = [D-tanφ*N]/A	(kN/m^2)	37

Table 5.7 Results of the limit equilibrium analysis for case 02, taking the width of the failure surface into account

Description	Formula	Unit	Value
Length	–	(m)	11.5
Width	–	(m)	6.5
Height	–	(m)	3.2
Contact area	A = L*B	(m^2)	74.75
Density	–	(kg/m^3)	2830
Slope angle	–	(°)	55
Assumed friction angle	–	(°)	35
Volume	V = l*w*h	(m^3)	239
Mass	M = ρ*V	(kg)	676,936
Force of gravity	G = m*g	(N)	6,640,742
Normal force	N = cosα*G	(N)	3,808,973
Driving force	D = sinα*G	(N)	5,439,777
Retaining force	R = tanφ*N	(N)	2,667,071
Factor of safety	**υ = R/D**	**–**	**0.5**
Minimum of cohesion for a labile stage (FS = 1)	c = [D-tanφ*N]/A	(kN/m^2)	37

assumed from the morphological north edge of the block to a steep inclined fault representing the south edge of the potential mass of failure. The results of the stability analysis without accounting for cohesion are provided in Table 5.7. The driving forces constitute 5440 kN, whereas the retaining forces are 2667 kN. The factor of safety can be specified as 0.5. The equal factor of safety like for case 1 results from the influence of the contact area respectively of volume on the acting forces. For a labile stage of the system we back-calculated the cohesion part assuming a factor of safety of 1, where the retaining force equals the driving force. The line at the bottom of Table 5.7 represents the necessary part of cohesion to keep the system at a labile stage, which could be assumed as 37 kN/m^2 for case 2.

5.1.6 Reconstruction of a Potential Mode of Failure

In a first step a potential mode of failure was analyzed reducing the stabilization problem to a simple 2D geometrical approach considering the block subjected to planar failure as the "active block" and the toe as the "passive block" (Fig. 5.4). The idea of this conceptual model is based on field data.

Excluding any deformation at the two blocks the geometric interpretation illustrates the increasing opening width due to the development of a rotational movement caused by the toe of the block. From stage 1 to stage 4 it can be followed that the contact area between the two blocks also decreases step by step. The geometric consideration provided the basis for the mechanical interpretation

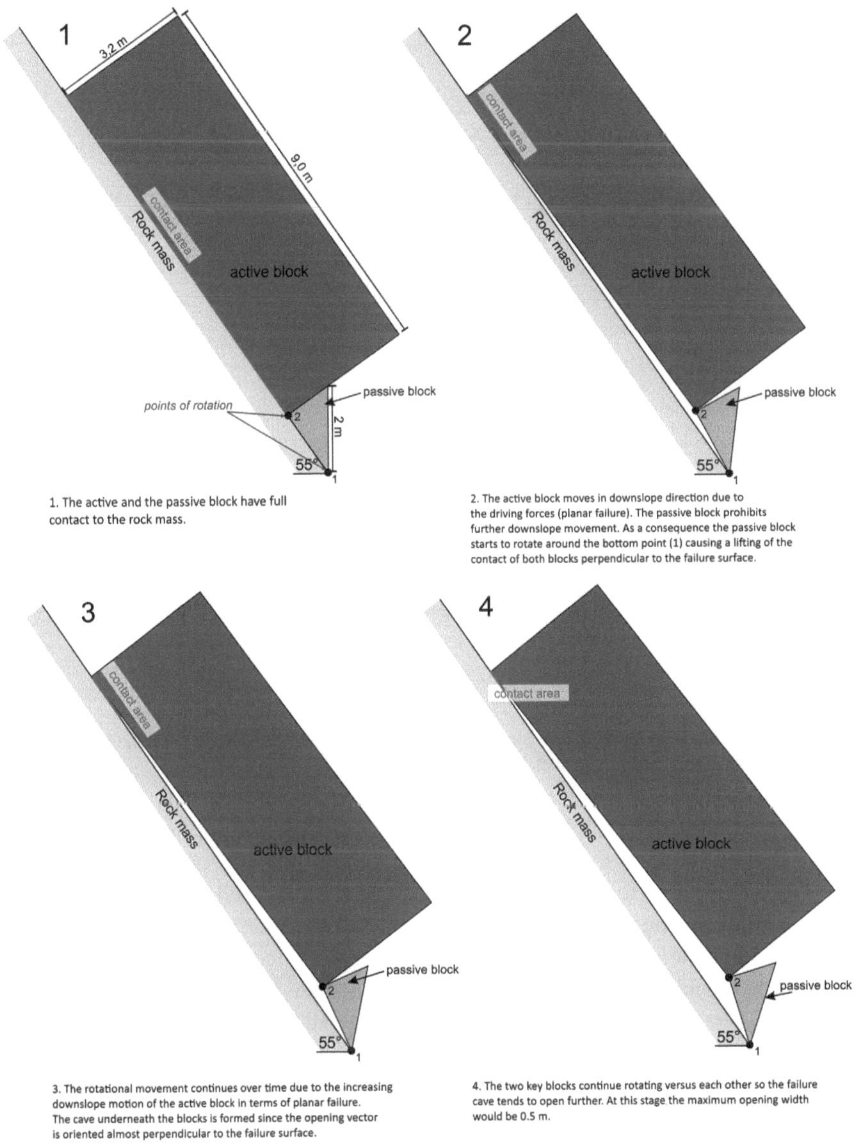

Fig. 5.4 Scaled sketch in 2D illustrating the way of failure reducing the setting to a 2D geometrical problem without taking deformation of the toe into account

including the approached block shapes and the observations concerning the failure surface. The decrease in the assumed opening width between the geometric and the mechanical interpretation could be explained by the observed deformation of the toe support. Due to the deformation of the toe, the active block is traced back to the failure surface.

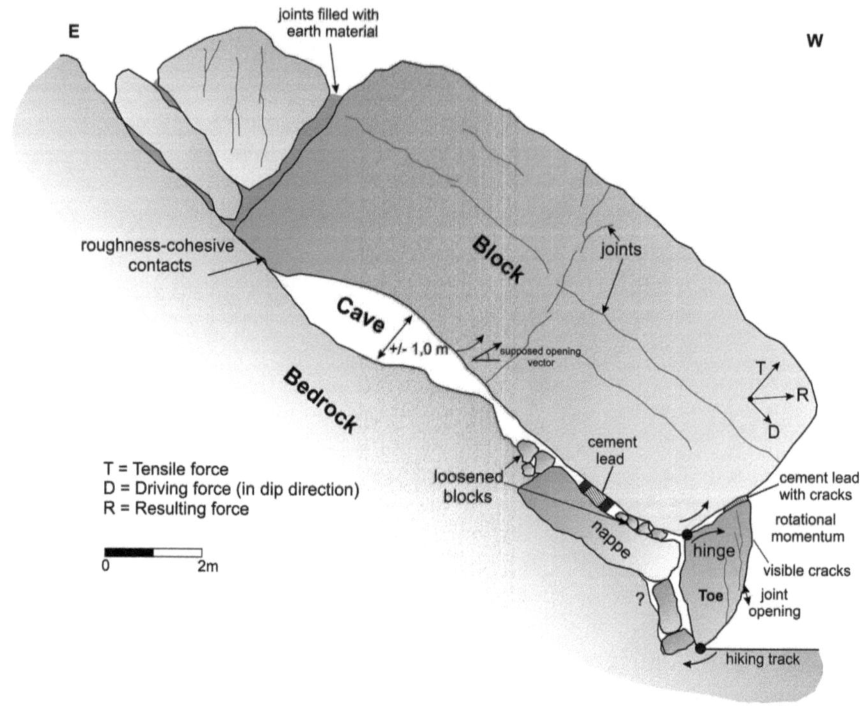

Fig. 5.5 Mechanical situation at the block. A rotational momentum at the toe of the block causes tensile stress affecting the block at the failure surface

The critical block at the Wachterl-Horn is limited by the following discontinuities: the lateral boundaries are represented by discontinuities belonging to joint set K3, the bottom and top boundaries are represented by bedding planes and the failure surface belongs to joint set K2, dipping westwards with 50–60° (Sect. 5.2.1).

The mechanical interpretation of the block suggests that the current case is not a simple type of planar failure, which can be characterized by the following features.

There is a toe-block at the bottom of the endangered block representing a resisting force on the one hand and acting as a hinge/rotational momentum on the other hand (Fig. 5.5). This rotational momentum seems to lead to a spreading of the block (from bottom to top), so that a cave underneath the block can be formed. Our findings from field work provide certain evidence for this hypothesis, since the cross sections through the failure cave suggest that the shape of the block and the rockmass mostly fit together like the pieces of a puzzle (Fig. 5.3). The cave shows an extensive opening width since spalling processes seem to occur in addition to the mechanical opening underneath the block. The spalling processes cause up to 0.4 m thick rock slabs crushing into the cave. The nappes and slabs show dimensions from 0.5 to 1.80 m in length and a maximum of 1 m in width (Fig. 5.3). Observing the blocks and rock nappes underneath the block, the amount and position of the material shut between block and bedrock changes from season to season indicating a dynamic system.

The upslope part of the contact block—bedrock is suggested to be characterized by a transition from cohesion to friction. If cohesive contacts in a strict sense are defined as rock bridges meaning an intact rock bond between block and bedrock, the rock bridges in the current case are at least fractured or already changed to rough rock-rock contacts. An evidence for this assumption can be adduced by field work, since grained rock flour is dispersed across the failure surface underneath the block in the rear part of the failure cave. The rock flour is assumed to result from sheared off roughness contacts partly connecting the block with the underlying rock mass. At the toe of the block, a fragmentation in terms of vertical cracking is highly visible. The toe-block seems to be divided into three separated blocks, each of them drawn through by horizontal and vertical cracks (opening widths 2–3 mm). The cracks provide certain evidence that at least an increased load is set as a burden on the toe. This leads to the assumption that the amount of intact rock bonds left in the upper part of the slope has to be decreased. We suppose that the main block affects the toe by providing an edge load at a very narrow contact area block-toe. The dimensions of the contact area between block and toe was assessed by placing a folding rule at the space in between both rock fragments.

5.1.7 Installation of Strain Gauges

For observing the movement rates of the block we installed four strain gauges at the failure surface of the critical block. The way of installation and usage of measuring device is explained in Sect. 4.1.7. The logger system was only installed for the winter season 2012–2013 (Fig. 5.6), due to the break-down of the system in May

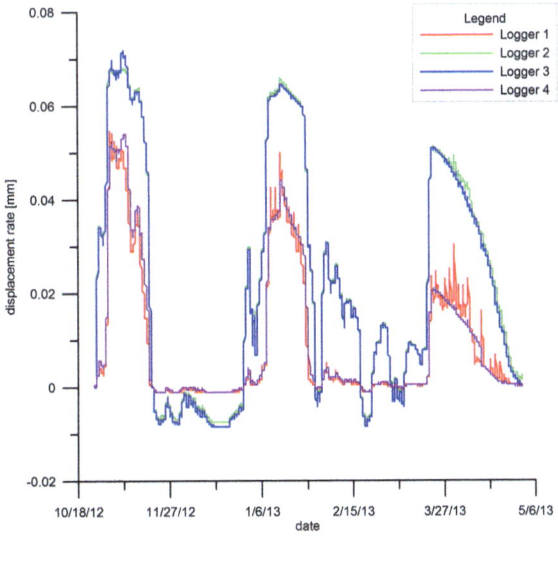

Fig. 5.6 Diagram of the data evaluation of the installed strain gauges. The data suggest a minor influence of the temperature changes on the recorded data

2013. The data illustrated in the plot in Fig. 5.6 show, that the displacement rates do not exceed values of 0.07 mm, varying between 0 and 0.07 mm. It is to note that “logger 1” corresponds to the logger at the outside of the critical block and “logger 4” is the strain gauge installed at the most inner part of the failure surface. The recording period lasts from October 2012 to May 2013.

5.2 Quantitative Magnitude Assessment

5.2.1 Scanline Analysis

For the current project site three scanlines were recorded: one “virtual” scanline to yield reconnaissance about the dominant joint set and as basic data for kinematic analysis; two further scanlines were recorded in the vicinity of the critical block at the Wachterl-Horn. The procedure of field recording and the way of evaluation are described in detail in Sect. 4.2.1.

In the following paragraph the results in terms of mean joint set orientations area described.

The results of the “virtual” scanline (190/00) show a tendency of three dominant joint sets in the carbonates of the Dachstein-Formation: the bedding (083/39), dipping eastwards and two joint sets K1 (251/53) and K2 (322/82); forming an almost orthogonal joint system. Figure 5.7 shows the contour plot of the “virtual” scanline including 105 recorded joint orientations, where the color scale stands for the density concentrations of the poles. The marked windows represent the area of joints taken into account for the respective joint set. The red window represents the bedding, the blue one K1 and the green one K2. The color code is the same for all three scanline evaluations. The yellow line represents the scanline orientation as a linear in the Schmidt Net. The mean joint set values determined using DIPS (Rocscience) are visualized in Table 5.8.

The scanline 01 (140/00) is situated in the south of the critical block at the Wachterl-Horn (Fig. 4.7). The results demonstrate similar results as the ones of the “virtual” scanline, suggesting three main joint sets specified as bedding (074/48), joint set K1 (243/59) and joint set K2 (337/82). Equally to the plot of the “virtual” scanline, Fig. 5.8 shows the defined joint sets marked by the colored windows in the contour plot. The mean joint set values determined using DIPS (Rocscience) are visualized in Table 5.9.

The scanline 02 (100/00; 130/00) is located in the direct surrounding of the critical block (Fig. 4.7). The evaluation shows, almost comparable to the other two, three dominating joint sets being named as the bedding (069/40), the joint set K1 (229/50) and the joint set K2 (339/87). Figure 5.9 represents the contour plot of scanline 02 in which the representative joint sets are marked with the colored window shapes. The yellow and green lines represent the orientations of the two scanline sections. The recorded joints were assigned to the respective scanline

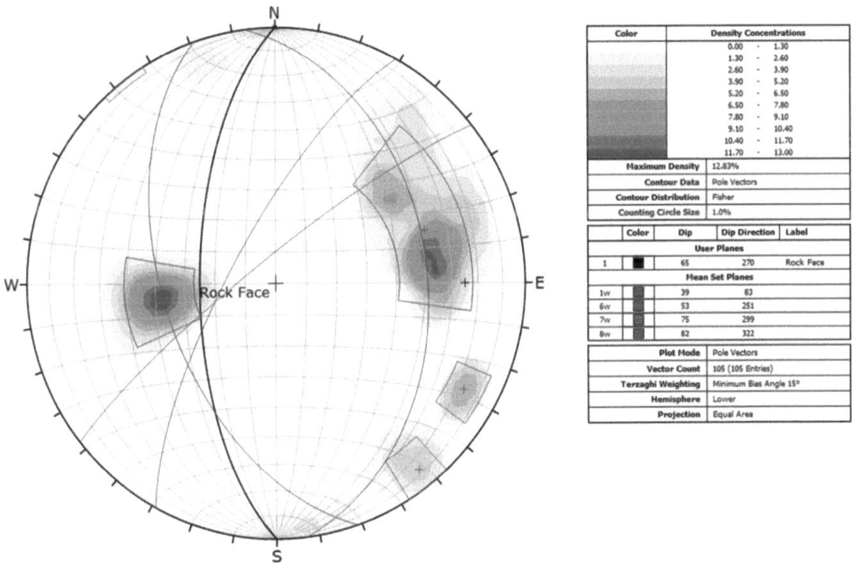

Fig. 5.7 Contour plot of the "virtual" scanline, which extends over a distance of approximately 700 m across the rockfall source area. The plot suggests three dominating discontinuity sets marked with the *coloured windows*

Table 5.8 Mean orientations of the joint sets resulting from the "virtual" scanline at the source area of the project site (105 total recorded discontinuities)

Discontinuity set	Dip direction	Dip
Bedding	083	39
Joint system K1	251	53
Joint system K2	299	75
Joint system K2	322	82

section for the Terzaghi weighting. The mean values for the determined joint sets are visualized in Table 5.10.

Figure 5.10 demonstrates the generalized discontinuity pattern related to the joint situation recorded at the release area. The color code of both figures is the same: the bedding is colored in red, joint set K1 in blue and joint set K2 in green. The results from scanline analysis are in agreement with the field observations, demonstrating that the discontinuities of the outcrop photo can be parallelized with the results of the joint pattern based on scanline analysis.

The results of the scanline analysis referring to Priest (1993) are described in the following paragraph. The scanline locations were selected due to the accessibility and outcrop conditions. Due to these two limiting factors it was possible to record two 20 m long scanline sections at the source area, of which the results are presented in Tables 5.11 and 5.12. The trace end is coded referring to the following categories: 1 = visible end in intact rock; 2 = visible trace end at another discontinuity; 3 = trace end not visible and 4 = cropping out. The curvature is coded

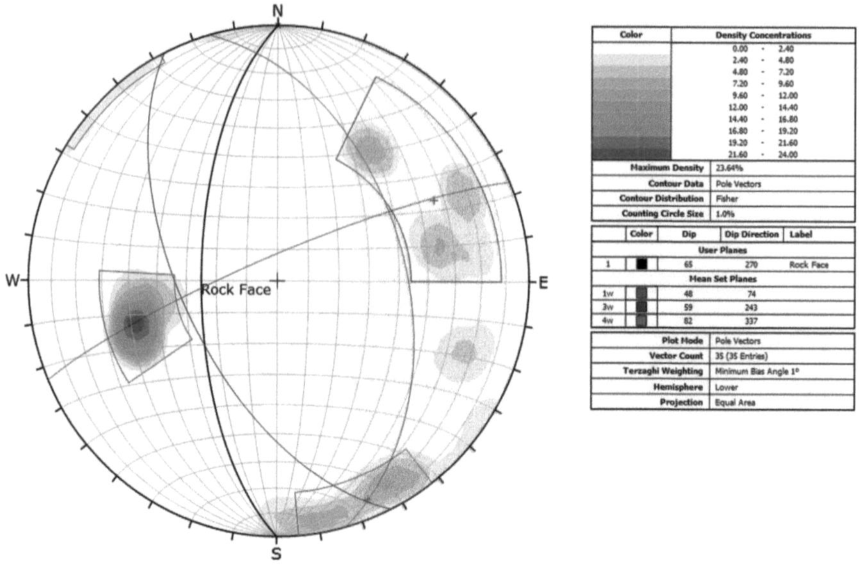

Fig. 5.8 Contour plot of scanline 01, which extends over a distance of 20 m and is located south of the critical block at the Wachterl-Horn. The plot suggests three dominating discontinuity sets marked with the *coloured windows*; the rock face is shown by the *black great circle*

Table 5.9 Mean orientations of the joint sets resulting from scanline 01 at the source area of the project site (35 total recorded discontinuities)

Discontinuity set	Dip direction	Dip
Bedding	074	48
Joint system K1	243	59
Joint system K2	337	82

according to 1 = straight and 5 = very curved. JRC indicated the joint roughness coefficient (qualitatively determined) and HTL means half trace length.

5.2.2 *Kinematic Analysis*

The details concerning the results of the joint recording and scanline analysis are described in Sect. 5.2.1.

The results from planar failure analysis (Fig. 5.11) show the rock face with a mean orientation of 270/65 and the associated Markland Plane colored in red, taking a friction angle of 35° into account. The pole cloud shaped with the blue window corresponds to joint set K1, with a mean orientation of 251/53 at the "virtual" scanline. The plot suggests that joint set K1 would be critical in terms of planar failure.

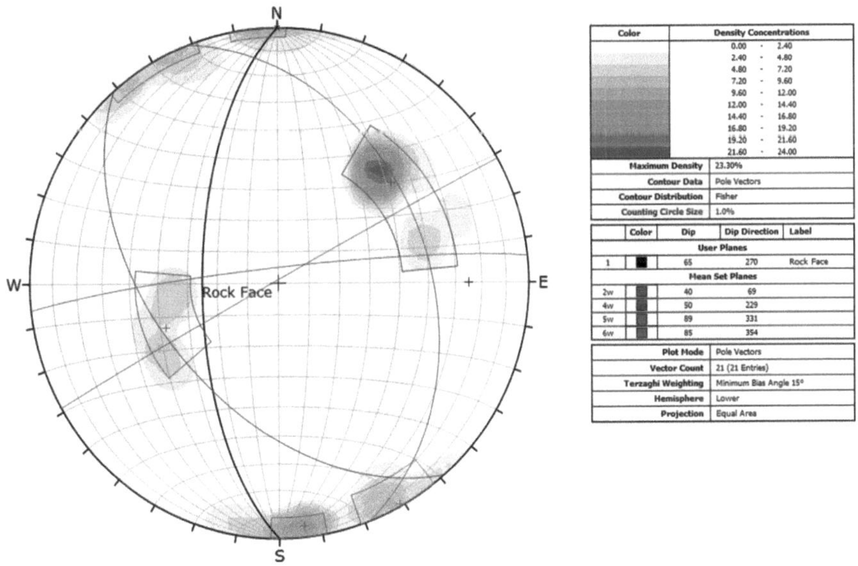

Fig. 5.9 Contour plot of scanline 02, which extends over a distance of 20 m and is located in the direct vicinity of critical block at the Wachterl-Horn. The scanline had to be divided into two sections due to morphological structures at the rock face. The plot suggests three dominating discontinuity sets marked with the *coloured windows*; the rock face is shown by the *black great circle*

Table 5.10 Mean orientations of the joint sets resulting from the scanline 02 at the source area of the project site (21 total recorded discontinuities)

Discontinuity set	Dip direction	Dip
Bedding	69	40
Joint system K1	229	50
Joint system K2	331	89
Joint system K2	354	85

Figure 5.12 represents the results from the wedge failure analysis under consideration of a friction angle of 35°. The black great circle represents the generalized slope orientation of 270/65. The red data points characterize the intersection-points of the dominating joint sets K1 and K2, which show potential for wedge failure in relation to the rock face. The results were filtered referring to the dominating critical joint sets, not taking the intersections of the bedding with joint set K1 respectively K2 into account. As a consequence the plot suggests that compared to all possible intersections of the 105 recorded values, approximately 20 % are endangered in terms of wedge failure.

The results of the direct toppling analysis show the critical intersections of the bedding and the joint set K2. The inner small circle represents the assumed friction angle of 35° whereas the outer small circle represents the inclination angle of the rock face. The results were filtered according to the impact on the direct toppling

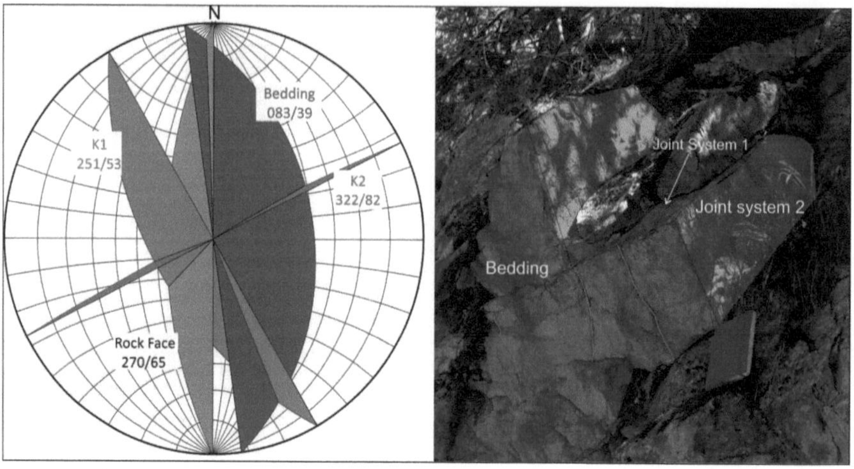

Fig. 5.10 *Left* Joint pattern of the discontinuity sets. The intersection of joint set K1/K2 could be critical for wedge failure. The joint set K1 shows potential for planar failure. Both cases can be verified in the field

mechanism. In total 167 intersection points for direct toppling are considered where 89.7 % belong to critical intersection points provided by the bedding and joint set K2. An additional sliding plane could be provided by joint set K1 (Fig. 5.13).

5.2.3 *Counting of Block Axes at the Talus Slope*

The recording of block axes at the talus slope included the measuring of the three block axes as well as the mean obstacle height (MOH). Due to the mean diameter being a common dimension specification for blocks, it was decided to present the mean block diameter as a characteristic value linked to the block volumes and the recorded MOH values as a parameter for the slope roughness for modelling with the code Rockyfor3D. The results of both parameters will be presented as value tables, whereas the plots will be presented in Sect. 6.2.3.

Table 5.13 represents the results of the mean block diameter distribution of all five sample areas in percent of the total amount of counted blocks. It is to be noted that only blocks with a diameter larger than 0.1 m were considered due to two facts: firstly very small blocks will be pressed into the slope material during the block-slope interaction and secondly, blocks with a mean diameter smaller than 0.1 m would contribute to slope roughness in an inferior role. The sample area 1 is situated nearly to the valley bottom, whereas the sample area 5 is situated near to the source area (Fig. 4.10). The results suggest that the main portion of the block sizes vary between 0.1 and 0.4 m in the mean diameter. For the sample areas 1, 2, 3 and 5 about 66–76 % of the block material is supposed to have a mean diameter of 0.1–0.3 m. Only sample area 4 contains 55 % in the diameter range of 0.1–0.3 m,

Table 5.11 Results of recording scanline 1 at the vicinity of the critical block

Distance of intersection (m)	Dip direction (°)	Dip (°)	HTL above scanline (m)	Trace end	JRC, 1–20	Curvature 1–5
0.80	286	75	2	1	6	1
0.95	290	70	15	3	5	1
1.90	80	38	0.6	1	8	2
2.20	216	53	10	1	4	2
2.60	295	68	2	1	10	3
2.70	328	78	2	1	10	3
3.28	320	90	1.5	1	11	2
4.20	338	80	10	3	12	1
5.05	310	90	>15	3	10	1
5.10	70	50	1.2	4	4	1
5.20	255	53	0.3	1	13	2
7.00	345	65	0.1	4	10	2
7.90	328	80	>15	3	11	2
8.25	326	80	2	1	14	3
8.70	245	70	2.5	1	15	2
9.70	78	48	2	1	16	1.0
10.25	332	75	2.5	1	14	4
11.15	330	85	>15	1	16	5
11.60	325	90	3	1	11	2
11.95	84	48	>15	2	15	1
13.08	305	90	>15	3	15	1
14.00	262	58	0	2	16	3
14.20	332	80	8	1	14	2
15.70	68	50	0.5	2	9	1
15.80	344	85	>15	3	14	1
16.20	300	85	10	1	9	2
17.70	262	70	2.5	2	17	3
17.70	343	85	10	2	8	3
17.90	360	75	1.2	2	13	2
18.10	358	85	1.5	2	12	1
18.22	350	85	1.5	2	12	1
18.40	355	85	1.5	2	14	1
19.00	70	53	0.5	1	12	1
19.40	290	60	>15	3	11	2
20.00	348	85	1.20	1	10	1

The scanline orientation is 140/00

but about 30 % in the diameter classes varying between 0.3 and 0.5 m. For all sample areas the part of large blocks diameters between 0.61 and 1.3 m decreases to 3.0–8.6 %.

Table 5.12 Results of recording scanline 2 at the vicinity of the critical block

Distance of intersection (m)	Dip direction (°)	Dip (°)	HTL above scanline (m)	Trace end	JRC von 1–20	Curvature 1–5
0.10	220	55	10	3	18	4
2.50	70	45	7	3	12	1
4.50	336	85	6	3	7	1
5.70	138	85	5	3	12	1
7.00	54	45	3	2	7	1
8.70	354	85	4	1	5	1
9.70	250	65	2.5	1	12	3
9.70	230	60	4.5	1	10	1
10.00	320	90	6	3	10	2
11.00	160	80	7	3	11	1
11.20	225	45	1	1	10	1
11.70	332	75	1.2	4	10	2
12.20	222	50	2	2	8	1
13.80	328	90	8	3	14	5
15.00	335	83	7	3	14	1
15.00	255	50	0.7	3	4	1
16.10	355	85	7	3	4	3
16.10	84	35	3	4	2	1
17.40	180	85	4	1	4	1
19.00	155	85	0.8	2	14	3

The scanline orientation is 100/00 for the first ten recorded meters and 130/00 for the intersection distance 10.00–20.00 m

Table 5.14 shows the distribution of the mean obstacle heights (MOH) measured in the 5 sample areas as percentage of the total recorded amount of blocks. For the recording only the blocks with a mean diameter larger than 0.1 m were considered analogue to the approach for the mean-diameter recording. The results of the MOH suggest that the dominant part of blocks in the sample areas 1 and 4 (about 77–84 %) vary between 0.01 and 0.2 m in MOH. In the sample areas 2, 3 and 5 the main part of the recorded blocks show MOHs between 0.01 and 0.1 m (about 77–86 %). Increasing MOHs in a range between 0.4 and 1.0 m are suggested to occur in the sample areas 1, 3, 4 and 5. The test area 1 contains 3.5 % of blocks providing a MOH between 0.5 and 1.0 m and area 4 contains about 7 % of blocks with a MOH varying between 0.4 and 1.0 m. For the areas 3 and 5 the amount of blocks with an increased MOH is less than 2 %. A discussion of the results of talus slope-recording in relation to the total number of considered blocks will be provided in Sect. 6.2.3.

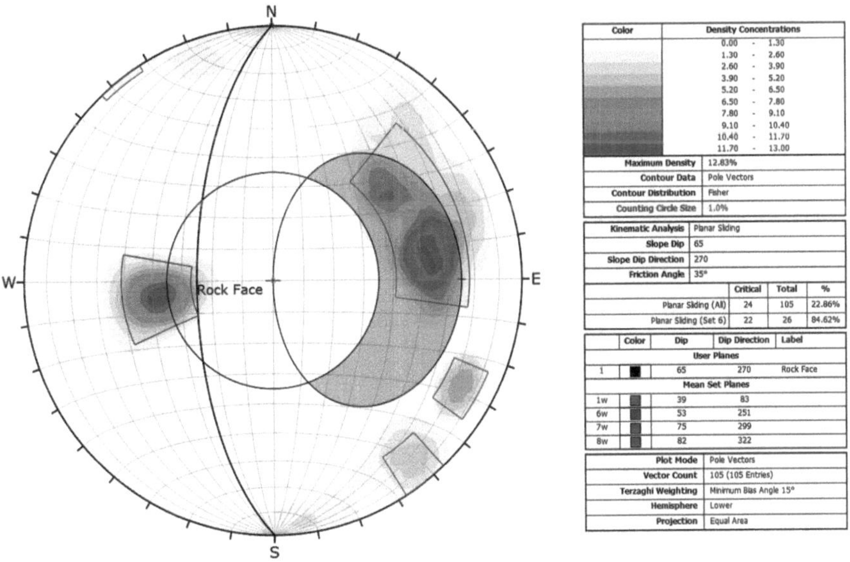

Fig. 5.11 Kinematic Analysis in terms of planar Sliding for a general orientation of the rock face of 270/65. The dominating joint sets are marked with the *coloured window* outlines. The discontinuities belonging to joint set K1 (*blue*) are predestined for planar sliding

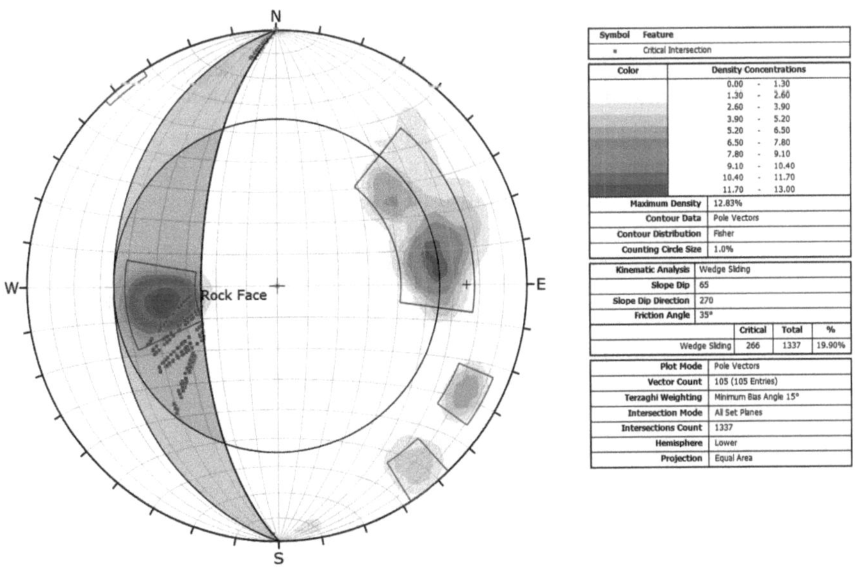

Fig. 5.12 Kinematic analysis for a general rock face orientation of 270/65 in terms of wedge sliding. The dominating joint sets are marked with the *coloured window* outlines. The data points mark the intersection locations of the joints sets K1 and K2 (*blue and green window outlines*).The black small circle represents a friction angle of 35°

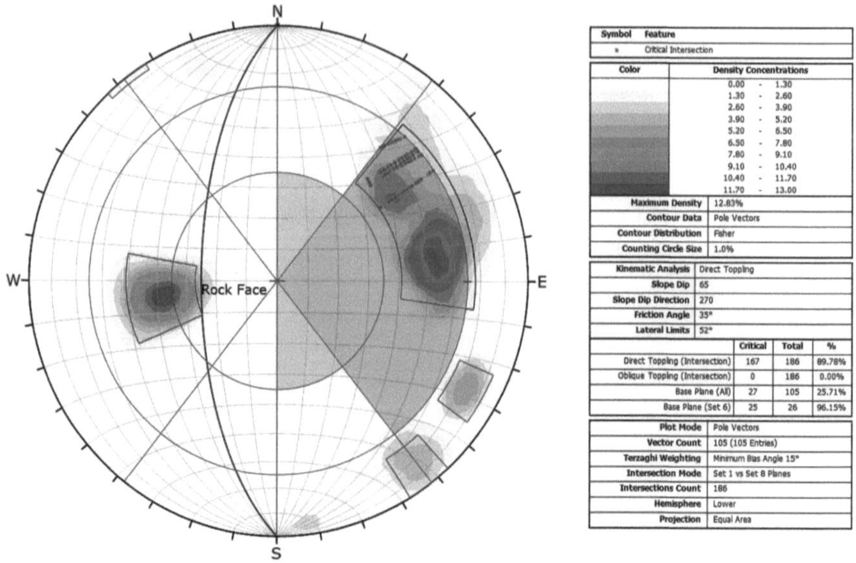

Fig. 5.13 Kinematic analysis of the generalized rock face 270/65 in terms of direct toppling. The red data points represent the intersection points of the bedding and joint set K2. These intersections would be responsible for the freeing of a certain block. The sliding plane could be provided by joint set K1. Inner small circle represents a friction angle of 35°, the outer small circle stands for the rock face dipping of 65°

Table 5.13 Results of the block recording at the talus slope showing the distribution of the mean block diameter (y-axis) as percentage of the total number of recorded blocks

Sample area	Mean diameter distribution given in % of the total amount of counted blocks					
	0.1–0.2 m	0.21–0.3 m	0.31–0.4 m	0.41–0.5 m	0.51–0.6 m	0.61–1.3 m
1	33.3	32.2	19.4	3.2	5.4	6.6
2	18.6	57.7	13.1	4.6	3.0	3.0
3	19.9	56.8	10.0	2.9	5.4	5.0
4	17.8	37.3	23.8	7.0	5.4	8.6
5	37.5	29.7	21.9	3.1	3.9	3.9

5.3 Rockfall Modelling

5.3.1 Rocky for 3D

Field Work

The following paragraph sums up the assessment of homogenous areas at the project site "Weißwand". The general process areas of the study site are shown in Fig. 5.14: The Carbonate rock cliffs at the top of the slope the densely forested

Table 5.14 Results showing the mean obstacle height distributions of the sample areas at the talus slope

Sample area	Mean obstacle height given in % of the total amount of counted blocks						
	0.01–0.05 m	0.051–0.1 m	0.11–0.2 m	0.21–0.3 m	0.31–0.4 m	0.41–0.5 m	0.51–1.0 m
1	22.1	33.7	27.9	11.6	1.2	0.0	3.5
2	65.6	21.3	8.3	3.2	1.6	0.0	0.0
3	57.3	20.8	14.6	3.5	2.3	1.2	0.4
4	18.9	26.8	30.5	11.6	5.5	4.3	2.4
5	39.8	41.0	10.8	6.0	1.2	1.2	0.0

The values are given as percentage of the total number of recorded blocks

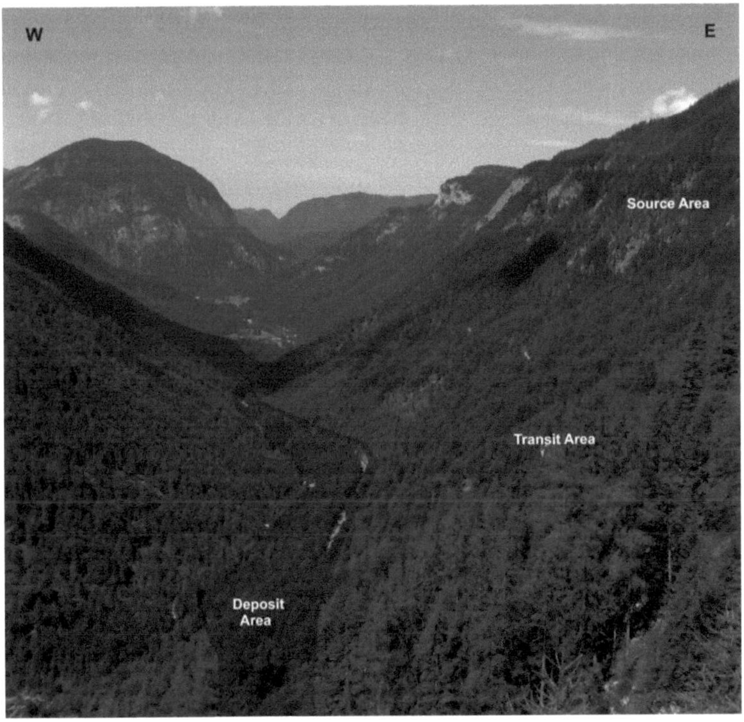

Fig. 5.14 Illustration of the main process areas at the Weißwand: The Carbonate rock cliffs at the *top*, the densely forested transit area with two forest roads and the deposit area at the *bottom* of the slope. The transit- and the deposit area can be mixed up with each other

transit area at the middle part down to the federal road and the deposit area down to the Schwarzbach at the valley bottom. The border between transit and deposit area was defined referring to the dominating process taking place in each area, but can be considered as fluent. The federal road is defined to be situated at the deposit area based on several events stopping at the road.

Table 5.15 Source area-parameters for the model set up of the Weißwand using Rockyfor3D

	D1 (m)	D2 (m)	D3 (m)	Rockdensity (kg/m^3)	Blockshape (–)
Min.	0.15	0.2	0.25	2830	1
Mean	0.2	0.25	0.35	2830	1
Max.	0.5	0.6	0.8	2830	1
Blocks	0.8	1.0	1.2	2830	1

The process areas were subdivided into homogenous areas defined referring to the process areas: the block subjected to be released (source area), the slope parameters (transit and deposit area) and the forest stand (whole slope).

The parameters for the source area are shown in Table 5.15. We decided to classify four volume classes based on the information obtained from quantitative magnitude assessment at the talus slope (Sect. 5.2.3).The class “Min.” represents the low magnitude rockfalls, which have an increased portion of the recorded blocks. The class “Mean” represents a middle diameter class in terms of volume-percentage as well as in terms of frequency. The class “Max.” stands for a low percentage, but increased block volumes. The class blocks was added on the one hand to include volumes of about 1 m^3 and on the other hand to compare the results with the hazard indication map of Bavaria suggesting block dimensions of 1.2 × 1.2 × 1.2 m for the Dachtein-Formation (Bayerisches Landesamt für Umwelt 2014: 14). The parameters of the slope are presented in Table 5.16.

GIS Set-up and Parameter Evaluation

The following paragraph describes the set-up of the model parameters in ArcGIS transferring the parameters recorded in the field into maps of homogenous areas for each parameter set.

The structure of the GIS Database is described in Sect. 4.3.2. For the process areas “Source area” and “Transit area” two maps of homogenous areas were created. The map of the source area (Fig. 5.15) provides information about the location of rock faces and thus of the source area. The feature class “source area” contains five field data types giving information about the block dimensions (d1, d2 and d3), the rock density and block shape (Table 5.15). The map of transit areas gives information about the underground parameters. This feature class contains four field data types: the roughness parameters (rg 70, rg 20 and rg 10) as the mean obstacle height in percentage of the homogenous area and the damping defined referring to the eight damping classes (Dorren 2012; Fig. 5.16).To account for the areas of windblow a second map was set up for the transit area, which incorporates the areas of windblow from the forest stand parameters (Fig. 5.17). In these areas the roughness was evidently increased by 20–30 cm due to root stems and root plates.

For characterizing the forest stand an additional map of homogenous areas has been created. The classes of homogenous areas are divided referring to the parameter classes of Tables 5.16 and 5.17. The feature class includes four field data types: the mean diameter at breast height (DBH mean), the standard deviation of the DBH (DBH std. dev.), the number of trees per hectare (nr_trees) and the percentage

Table 5.16 Illustration of the homogenous areas mapped at the Project site Weißwand

Exemplified photo	Damping	Roughness (MOH)	Short description
	6	Rg 10 = 0.1 Rg 20 = 0.05 Rg 70 = 0.03	Rockfall source area, partly with a decreasing slope angle than shown on this photo
	6	Rg 10 = 0 Rg 20 = 0 Rg 70 = 0	Smooth rock plane at the source area, for which no roughness was taken into account
	4	Rg 10 = 0.05 Rg 20 = 0.1 Rg 70 = 0.2	Area below the rock cliffs: gently dipping rock planes with vegetation cover
	4	Rg 10 = 0.1 Rg 20 = 0.3 Rg 70 = 0.2	Talus material with large blocks and root material providing an increased slope roughness. The talus material is covered by long grass

(continued)

Table 5.16 (continued)

Exemplified photo	Damping	Roughness (MOH)	Short description
	3	Rg 10 = 0.05 Rg 20 = 0.1 Rg 70 = 0.2	Typical impression from the talus material at the forested slope
	3	Rg 10 = 0.05 Rg 20 = 0.07 Rg 70 = 0.1	Outcrop in an avalanche trench. Due to the decreased block sizes the roughness was reduced
	3	Rg 10 = 0.05 Rg 20 = 0.07 Rg 70 = 0.1	Run-out area at the Schwarzbach, providing a decreasing slope angle and a light forest stand
	3/4	Rg 10 = 0.15 Rg 20 = 0.5 Rg 70 = 0.3	Areas of windblow: providing cut-off trees root stocks and root planes, leading to an increased slope roughness

(continued)

Table 5.16 (continued)

Exemplified photo	Damping	Roughness (MOH)	Short description
	0	Rg 10 = 100 Rg 20 = 100 Rg 70 = 100	Riverbed, Schwarzbach: the slope parameters for the riverbed were set according to the recommendation of the manual (Dorren 2012)
	7	Rg 10 = 0 Rg 20 = 0 Rg 70 = 0	Federal road B 305

Left Photo of the homogenous area; *middle* recorded damping parameter; *right* assessed MOH

of coniferous trees (conif_percent). The shapes of the homogenous areas of the forest stand are based on the information from aero photo analysis (20 cm resolution), validated during field work (Fig. 5.18).

Modelling

In this paragraph the results of selected rockfall scenarios are presented. The scenarios refer to the following parameter Matrix (Fig. 5.19). For each results paragraph the matrix will be prepended indicating the described scenarios in order to provide a structure to the reader.

Note

Since the project site Weißwand is densely forested all over the slope, the presented results are (except of the first presented scenario) all modelled taking the forest stand into account.

The following paragraph provides a glossary for the presented modelling results of Rockyfor3D:

- "Nr. Of deposited" blocks: number of block deposited in a certain raster cell (which is appropriate for hazard mapping of run-out zones)
- "E-mean": mean of the maximum kinetic energy values (translational and rotational) of the blocks, recommended for the generation of hazard maps)

Table 5.17 Homogenous areas for the forest stand parameters at the Weißwand

Exemplified photo	DBH mean/std. dev.	Stock density (number of trees/hectare)	Percent of coniferous trees
	30/14	280	80
	28/18	180	80
	23/11	25	90

- "Ph_95": confidence level of 95 % of the maximum jumping heights in m, measured perpendicular to the slope surface (recommended to be considered as the maximum jumping height)

The results of the quantitative recording of block volumes demonstrate that the dominating block diameters could be assigned to the block class "Min." and "Mean" (Sect. 5.2.3). For visualizing the run-out of the minimum block sizes, the scenario combining the underground class "Transit_1" and "Transit_2" with the block volume "Min" is chosen. The results of both underground-parameter classes are plotted in one figure.

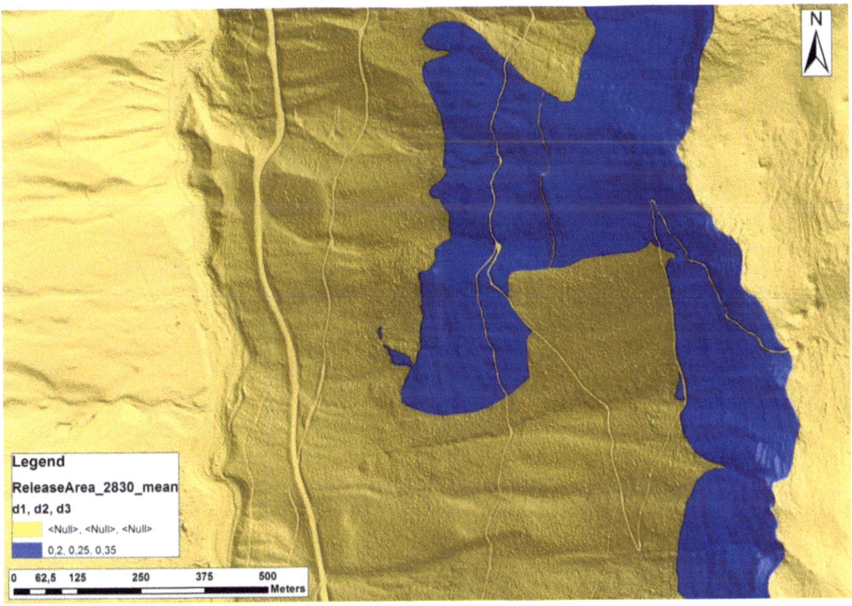

Fig. 5.15 Map of the release areas at the project site Weißwand. As an example the map of mean block sizes was chosen. The *blue* areas mark the rock faces here suggested to provide block sizes with the dimensions 0.2 × 0.25 × 0.35 m, a rock density of 2830 kg/m^3 and a block shape of 1 (Dorren 2012) (Color figure online)

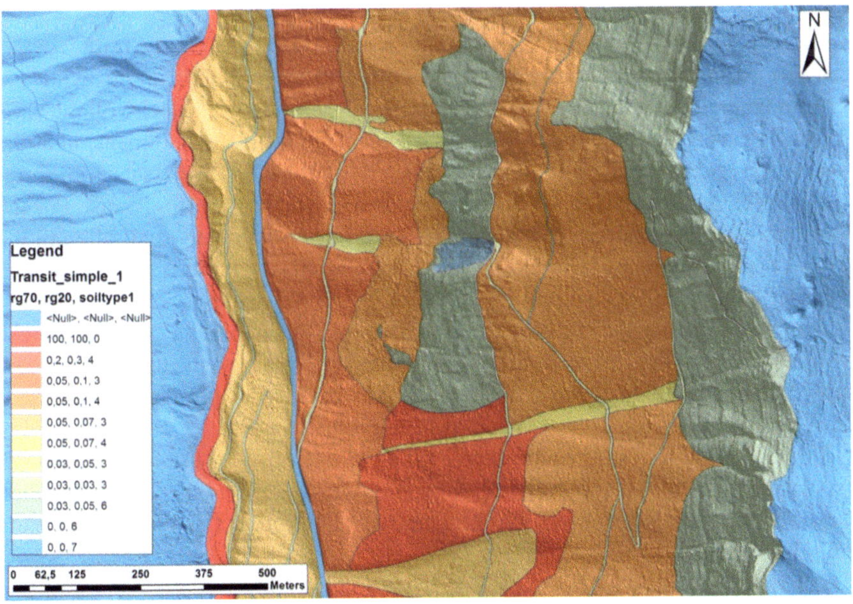

Fig. 5.16 Map of the transit and deposit area. As an example the map transit_1 was chosen, representing the underground parameters mapped in the field, validated referring to Dorren (2012). The legend shows the two dominating roughness classes of the MOH: rg70 and rg20. The *right column* shows the damping class

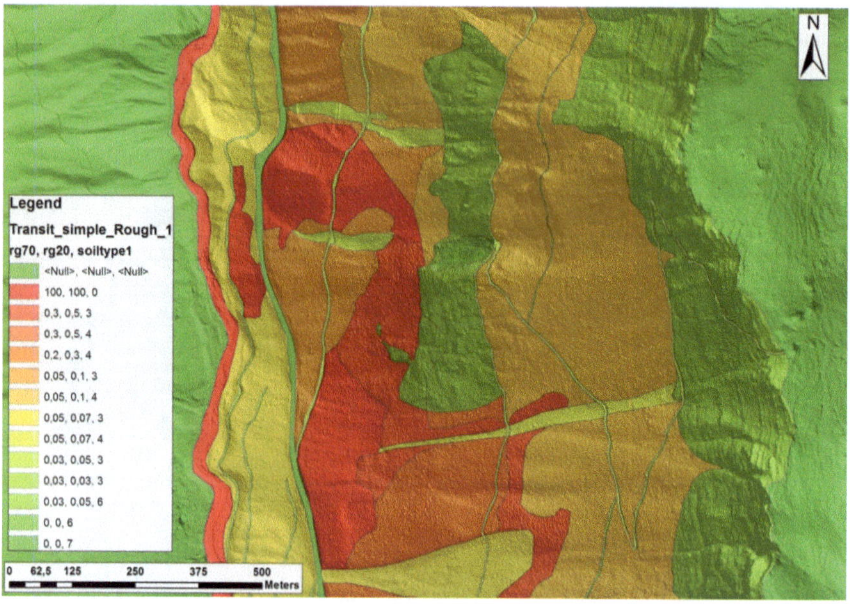

Fig. 5.17 Map of the transit and deposit area. As an example the map transit_1 was chosen, representing the underground parameters mapped in the field, validated referring to Dorren (2012). The legend shows the two dominating roughness classes of the MOH: rg70 and rg20. The *right column* shows the damping class. In addition the areas of windblow (orange colors with rg 70 values of 0.3) were taken into account based on the forest stand parameter set

Comparison of the Rockfall Run-Out with and Without Forest Stand

The following paragraph describes the variation of the rockfall run-out with and without taking the forest stand into account. For this purpose the block volume class "Mean" was selected due to the dominant percentage of this rock dimension at the talus slope (Sect. 5.2.3). The slope parameters of this modelling scenario correspond to the parameters recorded during field work (Transit_1) (Fig. 5.20). For the comparison of the results in consideration of the forest stand the visualisation of the "Nr. of deposited blocks" was chosen, which represents the number of accumulated rock per raster cell. The results presented in Fig. 5.21 suggest a reduced rockfall run-out when taking the forest stand into account, so the orange coloured blocks would travel less distance than the green coloured ones. Due to the talus slope being densely forested all over the project site it was decided to present the results of modelled scenarios under consideration of the forest stand.

In the following results section the modelling outcomes of the sensitivity analysis in terms of block volumes and slope roughness are presented. All described scenarios take the forest stand into account, since the project site is in large parts densely forested.

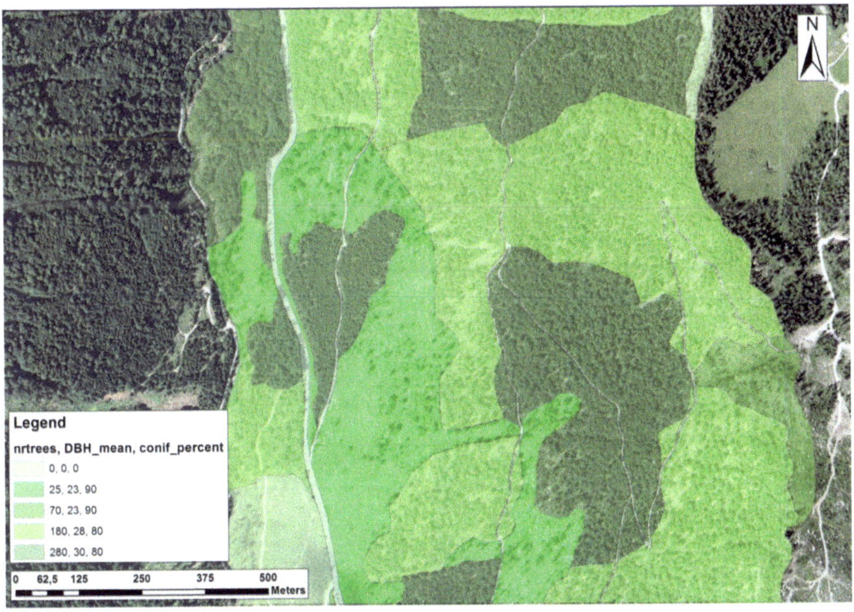

Fig. 5.18 Map of the forest stand parameters at the Weißwand project site. The parameters listed in the legend show in the *left column* the number of trees per hectare, in the *middle* the mean diameter at breast height (DBH_mean) and in the *right column* the percent of coniferous trees (Dorren 2012)

block volume / underground parameters	Min. 0.15x 0.2x 0.25 m	Mean 0.2x 0.25x 0.35 m	Max. 0.5x 0.6x 0.8 m	blocks 0.8x 1.0x 1.2 m
Transit_1 parameters based on fieldwork; validated ref. to Dorren (2010)				
Transit_2 increasing roughness (MOH) + 0.1 m at the talus slope + 0.15 m at the valley bottom				
Transit_rough_1 parameters based on fieldwork; validated ref. to Dorren (2010) incorporating areas of windbreakage				
Transit_rough_2 increasing roughness (MOH) + 0.1 m at the talus slope + 0.15 m at the valley bottom incorporating areas of windbreakage				

Fig. 5.19 Parameter Matrix for the rockfall modelling using the code Rockyfor 3D. Four classes of block volumes based on quantitative field recording were combined with four ground parameter-classes

underground parameters \ block volume	Min. 0.15x 0.2x 0.25 m	Mean 0.2x 0.25x 0.35 m	Max. 0.5x 0.6x 0.8 m	Blocks 0.8x 1.0x 1.2 m
Transit_1 parameters based on fieldwork; validated ref. to Dorren (2010)		X		
Transit_2 increasing roughness (MOH) + 0.1 m at the talus slope + 0.15 m at the valley bottom				
Transit_rough_1 parameters based on fieldwork; validated ref. to Dorren (2010) incorporating areas of windbreakage				
Transit_rough_2 increasing roughness (MOH) + 0.1 m at the talus slope + 0.15 m at the valley bottom incorporating areas of windbreakage				

Fig. 5.20 Parameter Matrix showing the chosen scenario: Mean/Transit_1 with and without taking the forest stand into account

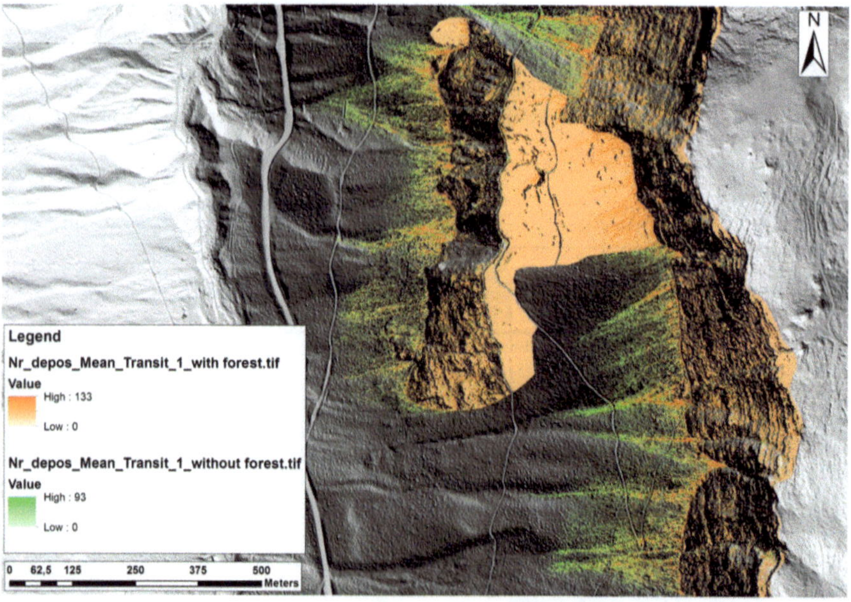

Fig. 5.21 Comparison of the rockfall run-out with and without taking the forest stand into account for the block volume class "Mean"; visualized as the Nr. of deposited rocks

Minimum Block Sizes (0.15 × 0.2 × 0.25 m)

Figure 5.23 shows the number of deposited blocks for the mentioned scenarios of Fig. 5.22. The blocks of the assumed minimum volume would not reach the federal road, neither in case of the underground parameters "Transit_1" nor in case of

underground parameters / block volume	Min. 0.15x 0.2x 0.25 m	Mean 0.2x 0.25x 0.35 m	Max. 0.5x 0.6x 0.8 m	Blocks 0.8x 1.0x 1.2 m
Transit_1 parameters based on fieldwork; validated ref. to Dorren (2010)	X			
Transit_2 increasing roughness (MOH) + 0.1 m at the talus slope + 0.15 m at the valley bottom	X			
Transit_rough_1 parameters based on fieldwork; validated ref. to Dorren (2010) incorporating areas of windbreakage				
Transit_rough_2 increasing roughness (MOH) + 0.1 m at the talus slope + 0.15 m at the valley bottom incorporating areas of windbreakage				

Fig. 5.22 Parameter Matrix showing the chosen scenario: Min/Transit_1 and Transit_2

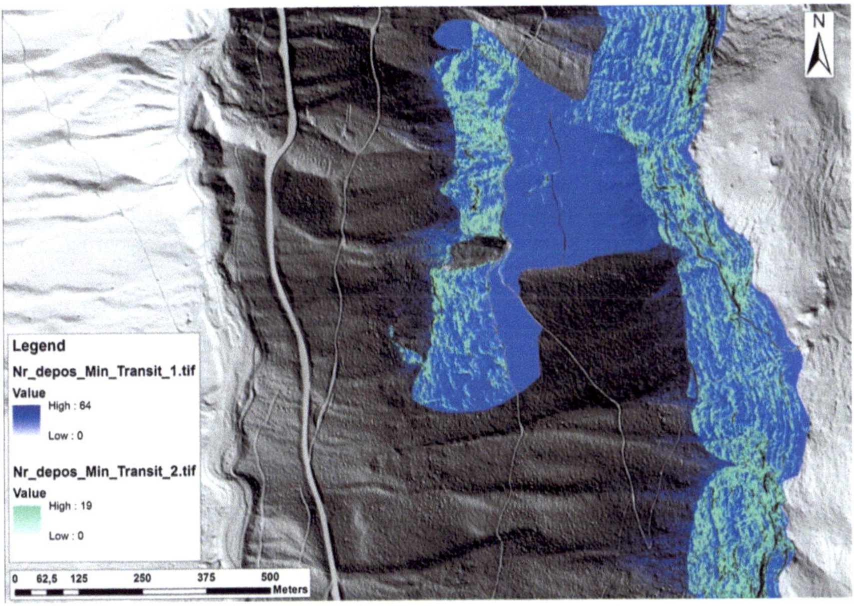

Fig. 5.23 Visualization of the number of deposited blocks for the scenario block volume "Min" combined with the underground classes "Transit_1" and "Transit_2" (Fig. 5.22). The blocks (0.15 × 0.2 × 25 m) will not reach the federal road in both cases of underground parameters

"Transit_2". The run-out zone of blocks following the avalanche trenches (scenario "Transit 1") also end evidently above the federal road. The more gentle inclined area between the steep rock faces seems to act as the main deposit area in case of the smaller block sizes. Since the focus of the project is set on the hazard assessment of the federal road at the bottom of the slope, the modelling results presented in the following sections, focus on the block volumes "Mean", "Max." and "Blocks", as the increased cubatures.

Mean Block Sizes (0.2 × 0.25 × 0.35 m)

For the described rockfall scenario the block size "Mean" was combined with the underground parameter classes of "Transit_1" and "Transit_2", where "Transit_2" stands for an increased roughness at the talus slope (Fig. 5.24). The rockfall run-out zones (nr. of deposited blocks) for scenario "Transit_1" (orange color) show a tendency to follow the avalanche trenches down the slope, but do in far areas not reach the federal road B 305; except in case of large avalanche trench in the north of the project site (Fig. 5.25).The run-out zones of "Transit_2" do not descend far down the talus slope, but mainly stop at the source area, due to the increased roughness of this parameter set. The maximum block amount deposited in a certain raster cell can be specified with 133 blocks in case of "Transit_1" and 25 blocks in case of "Transit_2". The difference in the maximum block amount per raster cell is caused due to the varying size of the deposit area in both cases.

The trajectories of the mean kinetic energies suggest that for scenario "Transit_2" only the blocks descending from the steep rockfaces travel far enough to result in recordable trajectories (Fig. 5.26). All Trajectories from this scenario end at the talus slope above the federal road. In case of scenario "Transit_1" trajectories at the middle of the project site show that the blocks stop very near to the road. At the north part the trajectories of two blocks cross the road. In case of "Transit_2" none of the kinetic energy trajectories reaches the federal road. The maximum values of the mean kinetic energies are 58 kJ in case of scenario

underground parameters / block volume	Min. 0.15x 0.2x 0.25 m	Mean 0.2x 0.25x 0.35 m	Max. 0.5x 0.6x 0.8 m	Blocks 0.8x 1.0x 1.2 m
Transit_1 parameters based on fieldwork; validated ref. to Dorren (2010)		X		
Transit_2 increasing roughness (MOH) + 0.1 m at the talus slope + 0.15 m at the valley bottom		X		
Transit_rough_1 parameters based on fieldwork; validated ref. to Dorren (2010) incorporating areas of windbreakage				
Transit_rough_2 increasing roughness (MOH) + 0.1 m at the talus slope + 0.15 m at the valley bottom incorporating areas of windbreakage				

Fig. 5.24 Parameter Matrix showing the chosen scenario: the block sizes of the class "Mean" (0.2 × 0.25 × 0.35 m) were combined with the underground classes "Transit_1" and "Transit_2", where "Transit_2" is the parameter set with increased roughness

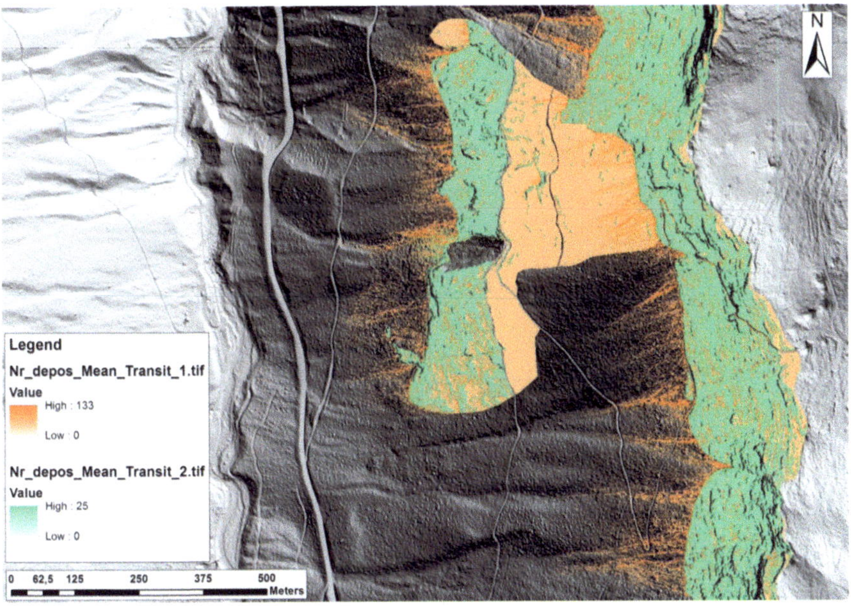

Fig. 5.25 Number of deposited blocks for the block size "Mean" (0.2 × 0.25 × 0.35 m) combined with the underground parameters of the scenarios "Transit_1" and "Transit_2". The run-out zones of "Transit_1" are illustrated in *orange colors*, the ones of "Transit_2" in *turquoise colors*

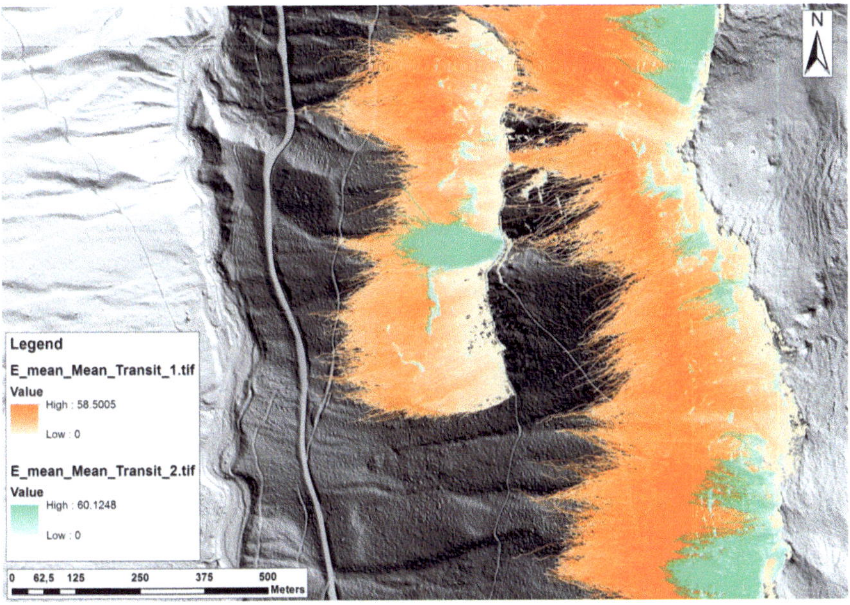

Fig. 5.26 Mean kinetic energies of the blocks belonging to the class "Mean" (0.2 × 0.25 × 0.35 m), combined with the underground parameters of the scenarios "Transit_1" and "Transit_2". The run-out zones of scenario "Transit_1" are illustrated in *orange colors*, the ones of "Transit_2" in *turquoise colors*

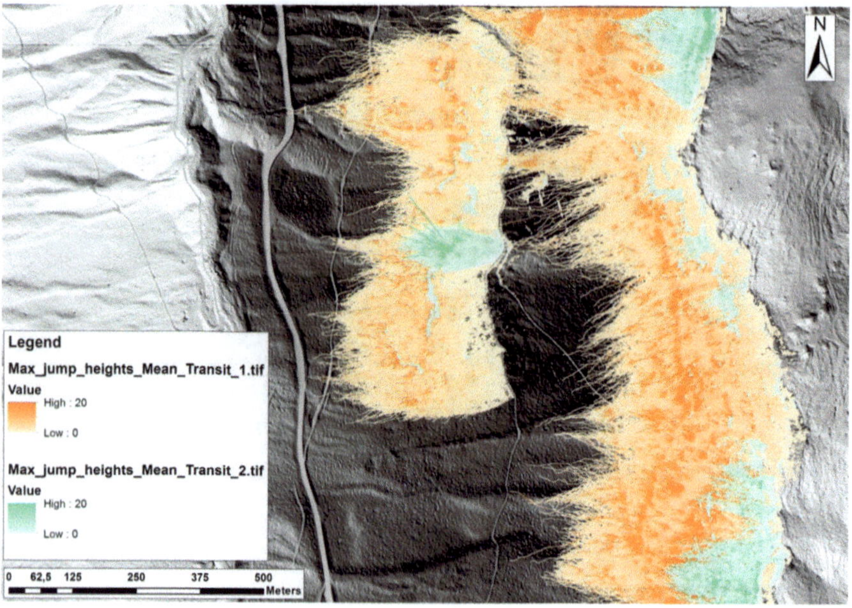

Fig. 5.27 Maximum jumping heights of the blocks belonging to the class "Mean" (0.2 × 0.25 × 0.35 m), combined with the underground parameters of the scenarios "Transit_1" and "Transit_2". The run-out zones of scenario "Transit_1" are illustrated in *orange colors*, the ones of "Transit_2" in *turquoise colors*

"Transit_1" and 60 kJ in case of scenario "Transit_2" for an assumed block size of 0.2 × 0.25 × 0.35 m. The mean kinetic energies include the rotational and the translational energy of the blocks.

The maximum jumping height for both scenarios, "Transit_1" and "Transit_2" can be specified as 20 m (Fig. 5.27). This maximum jumping height is measured normal to the slope surface. This constitutes why the maximum jumping height mainly occur at steep rock face steps or at forest roads lined with a rock step up slope. In large extends the jumping height can be considered as 3–10 m above the slope surface for both modelled scenarios in case of block class "Mean".

Max Block Sizes (0.5 × 0.6 0.8 m)

For the described rockfall scenario the block size "Max." was combined with the underground parameter classes of "Transit_1" and "Transit_2", where "Transit_2" stands for an increased roughness at the talus slope (Fig. 5.28). The rockfall run-out zones (nr. of deposited blocks) for scenario "Transit_1" (purple color) mainly follow the avalanche trenches down the slope (Fig. 5.29). Due to the morphological structures "hot spots" of rockfalls arise at the federal road at the apertures of the avalanche trenches. The run-out zones of "Transit_2" do not descend that far down the talus slope and mainly stop at the source area and the talus slope due to the increased roughness of this parameter set. The absolute maximum block amount

block volume / underground parameters	Min. 0.15x 0.2x 0.25 m	Mean 0.2x 0.25x 0.35 m	Max. 0.5x 0.6x 0.8 m	Blocks 0.8x 1.0x 1.2 m
Transit_1 parameters based on fieldwork; validated ref. to Dorren (2010)			X	
Transit_2 increasing roughness (MOH) + 0.1 m at the talus slope + 0.15 m at the valley bottom			X	
Transit_rough_1 parameters based on fieldwork; validated ref. to Dorren (2010) incorporating areas of windbreakage				
Transit_rough_2 increasing roughness (MOH) + 0.1 m at the talus slope + 0.15 m at the valley bottom incorporating areas of windbreakage				

Fig. 5.28 Parameter Matrix showing the chosen scenario: the block sizes of the class "Max."(0.5 × 0.6 × 0.8 m) were combined with the underground classes "Transit_1" and "Transit_2", where "Transit_2" is the parameter set with increased roughness

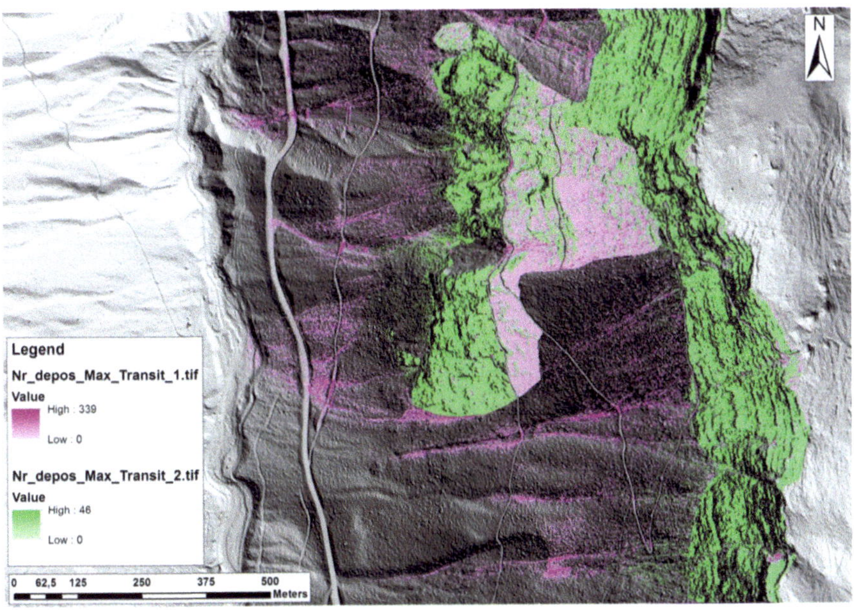

Fig. 5.29 Number of deposited blocks for the block size "Max." (0.5 × 0.6 × 0.8 m) combined with the underground parameters of the scenarios "Transit_1" and "Transit_2". The run-out zones of "Transit_1" are illustrated in *purple colors*, the ones of "Transit_2" in *green colors*

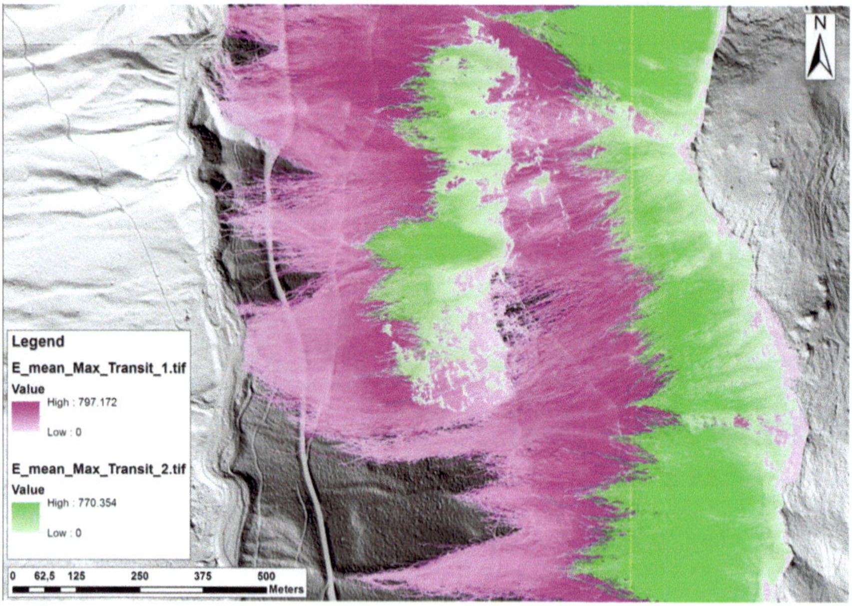

Fig. 5.30 Mean kinetic energies of the blocks belonging to the class "Max." (0.5 × 0.6 × 0.8 m), combined with the underground parameters of the scenarios "Transit_1" and "Transit_2". The run-out zones of scenario "Transit_1" are illustrated in *purple colors*, the ones of "Transit_2" in *green colors*

deposited in a certain raster cell can be specified with 339 blocks in case of "Transit_1" and 46 blocks in case of "Transit_2". The most raster cells for scenario "Transit_1" indicate block amounts of 10 to 30 blocks per raster cell. In case of scenario "Transit_2" most raster cells indicate a deposited block amount of 5 to 10 blocks per raster cell. The difference in the maximum amount of blocks per raster cell is caused due to the varying size of the deposit area in both cases.

The kinetic energy trajectories from scenario "Transit_1" suggest that the most blocks cross the federal road (Fig. 5.30). At the south part of the project site, where rockfall material descends only from the top of the slope, the rocks are not supposed to reach the road section. All descending rocks from scenario "Transit_2" stop at the talus slope above the federal road. In case of "Transit_2" none of the kinetic energy trajectories reaches the federal road. The maximum values of the mean kinetic energies are approximately 800 kJ in case of scenario "Transit_1" and 770 kJ in case of scenario "Transit_2" for an assumed block size of 0.5 × 0.6 × 0.8 m. The mean kinetic energies include the rotational and the translational energy of the blocks.

The maximum jumping height for both scenarios, "Transit_1" and "Transit_2" can be stated as 20 m (Fig. 5.31). This maximum jumping height is defined normal to the slope surface. This constitutes why the maximum jumping height mainly occur at steep rock face steps or at forest roads lined with a rock step up- or down

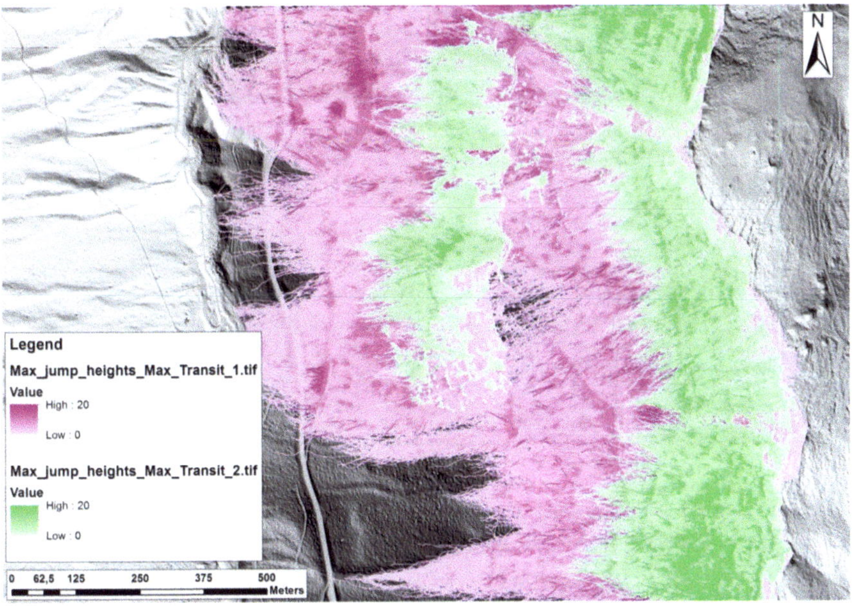

Fig. 5.31 Maximum jumping heights of the blocks belonging to the class "Max." (0.5 × 0.6 × 0.8 m), combined with the underground parameters of the scenarios "Transit_1" and "Transit_2". The run-out zones of scenario "Transit_1" are illustrated in *purple colors*, the ones of "Transit_2" in *green colors*

slope. In large extends the jumping height can be considered as 3–10 m above the slope surface for both modelled scenarios in case of block class "Max".

Rock Class "Blocks" (0.8 × 1.0 × 1.2 m)

For the described rockfall scenario the block size "Blocks" was combined with the underground parameter classes of "Transit_1" and "Transit_2", where "Transit_2" stands for an increased roughness at the talus slope (Fig. 5.32). The rockfall run-out zones (nr. of deposited blocks) for scenario "Transit_1" (blue color) mainly follow the avalanche trenches down the slope (Fig. 5.33). Due to the morphological structures "hot spots" of rockfalls arise at the federal road at the apertures of the avalanche trenches. An evident portion of blocks reaches the Schwarzbach at the valley bottom. The run-out zones of "Transit_2" (red color) do not descend that far down the talus slope and mostly stop at the source area and the talus slope due to the increased roughness of this parameter set. The absolute maximum block amount deposited in a certain raster cell can be specified with 410 blocks in case of "Transit_1" and 309 blocks in case of "Transit_2". The most raster cells for scenario "Transit_1" indicate block amounts of 5 to 20 blocks per raster cell at the talus slope and 20 to 60 blocks per cell at the deposit areas at the federal road. In case of scenario "Transit_2" most raster cells indicate a deposited block amount of 3 to 15

underground parameters \ block volume	Min. 0.15x 0.2x 0.25 m	Mean 0.2x 0.25x 0.35 m	Max. 0.5x 0.6x 0.8 m	Blocks 0.8x 1.0x 1.2 m
Transit_1 parameters based on fieldwork; validated ref. to Dorren (2010)				X
Transit_2 increasing roughness (MOH) + 0.1 m at the talus slope + 0.15 m at the valley bottom				X
Transit_rough_1 parameters based on fieldwork; validated ref. to Dorren (2010) incorporating areas of windbreakage				
Transit_rough_2 increasing roughness (MOH) + 0.1 m at the talus slope + 0.15 m at the valley bottom incorporating areas of windbreakage				

Fig. 5.32 Parameter Matrix showing the chosen scenario: the block sizes of the class "Blocks" (0.8 × 1.0 × 1.2 m) were combined with the underground classes "Transit_1" and "Transit_2", where "Transit_2" is the parameter set with increased roughness

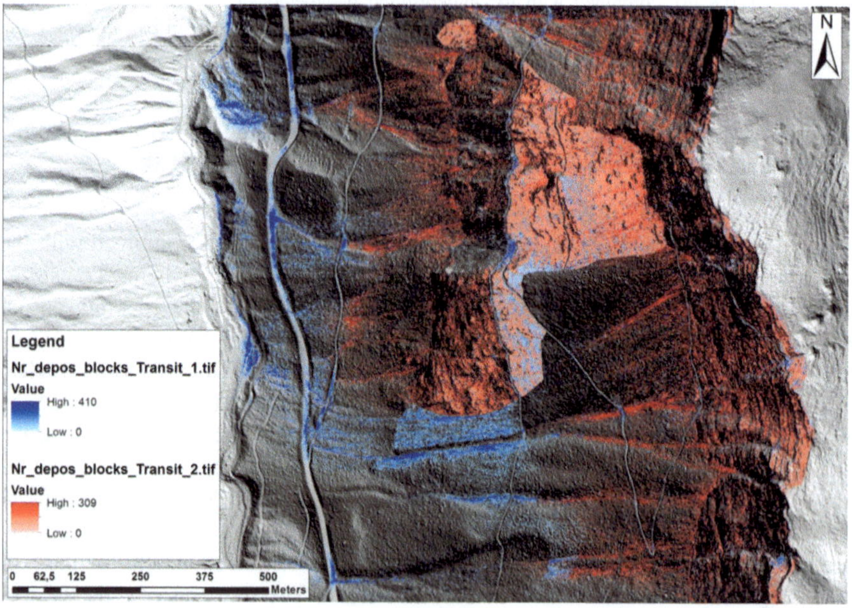

Fig. 5.33 Number of deposited blocks for the block size "Blocks" (0.8 × 1.0 × 1.2 m) combined with the underground parameters of the scenarios "Transit_1" and "Transit_2". The run-out zones of "Transit_1" are illustrated in *blue colors*, the ones of "Transit_2" in *red colors*

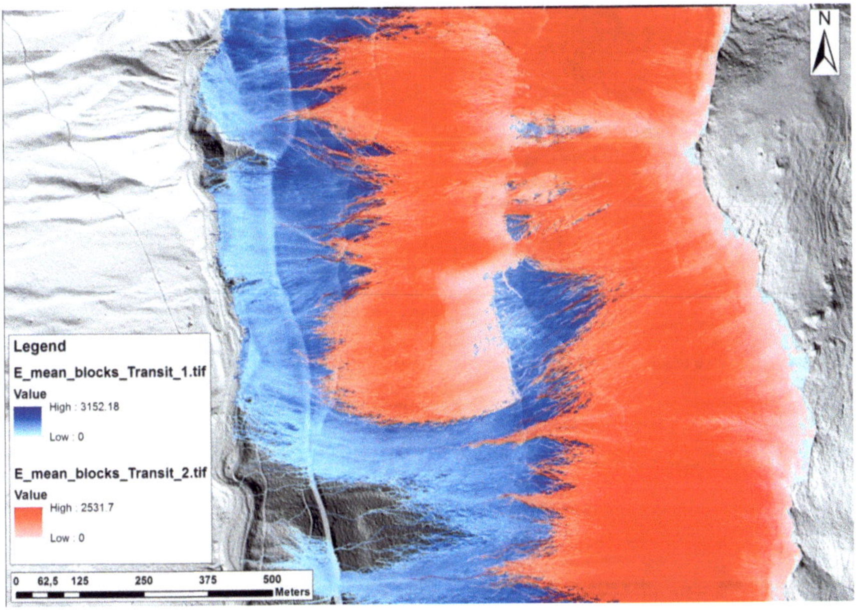

Fig. 5.34 Mean kinetic energies of the blocks belonging to the class "Blocks" (0.8 × 1.0 × 1.2 m), combined with the underground parameters of the scenarios "Transit_1" and "Transit_2". The run-out zones of scenario "Transit_1" are illustrated in *blue colors*, the ones of "Transit_2" in *red colors*

blocks per raster cell. The difference in the maximum amount of blocks per raster cell may be caused due to the varying size of the deposit area in both cases.

The kinetic energy trajectories from scenario "Transit_1" suggest that the most blocks cross the federal road (Fig. 5.34). Most descending rocks from scenario "Transit_2" stop at the talus slope above the federal road, but the trajectories at the middle and north part of the project site tend to cross the federal road. The maximum values of the mean kinetic energies are approximately 3150 kJ in case of scenario "Transit_1" and approximately 2530 kJ in case of scenario "Transit_2" for an assumed block size of 0.8 × 1.0 × 1.2 m. The mean kinetic energies include the rotational and the translational energy of the blocks.

The maximum jumping height for both scenarios, "Transit_1" and "Transit_2" can be stated as 20 m (Fig. 5.35). This maximum jumping height is defined normal to the slope surface. This constitutes why the maximum jumping height mainly occur at steep rock face steps or at forest roads lined with a rock step up- or down slope. In large extends the jumping height can be considered as 3–10 m above the slope surface for both modelled scenarios in case of block class "Max".

Rock Class "Blocks" (0.8 × 1.0 × 1.2 m) Including Areas of windblow

The windblow areas at the project site Weißwand result from the thunderstorm named "Kyrill", which destroyed large forest areas in January 2007. The extension

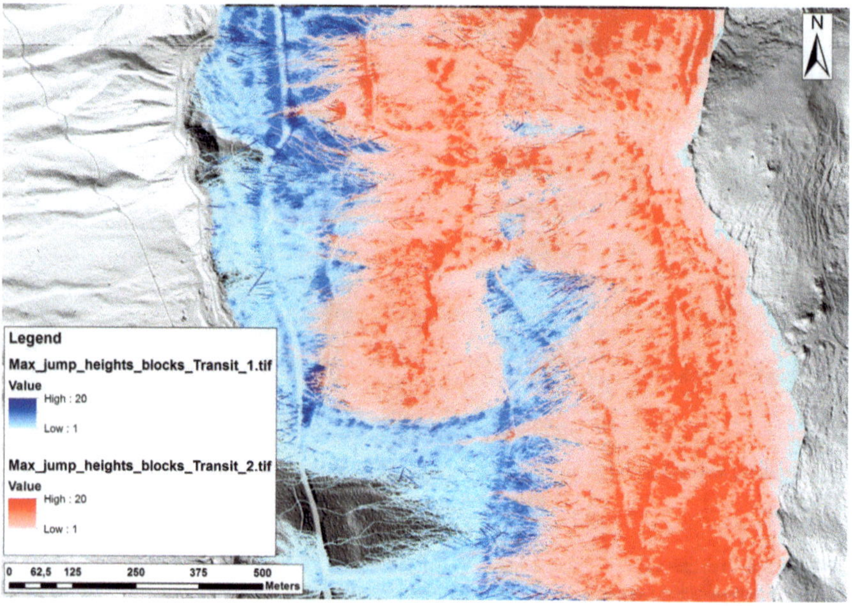

Fig. 5.35 Maximum jumping heights of the blocks belonging to the class "Blocks" (0.8 × 1.0 × 1.2 m), combined with the underground parameters of the scenarios "Transit_1" and "Transit_2". The run-out zones of scenario "Transit_1" are illustrated in *blue colors*, the ones of "Transit_2" in *red colors*

of the windblow area is based on the forest stand and transit-parameter set (Fig. 5.36) and was mapped according to aerophotos (Fig. 5.37). The increased roughness of the tree- and root stems evidently influences the run-out of blocks. To show the influence of an area of windblow the scenario of block class "blocks" was chosen, since the effect of increasing roughness is strongly visible for this block size. The scenario "Transit_rough_1" is based on the underground parameters mapped in the field and validated referring to Dorren (2012).

In Case of scenario "Transit_1" (blue colors, Fig. 5.38) the hot spots of deposited blocks at the federal road correlates with the apertures of avalanche trenches and most of the blocks with the dimensions of 0.8 × 1.0 × 1.2 m reach the road. The absolute maximum of deposited blocks per raster cell is 410 for this scenario. The mean range of deposited blocks per raster cell could be specified as 10 to 40 at the talus slope and 30 to approximately 100 at the valley bottom. In contrast Scenario "Transit_rough_1" (red colors, Fig. 5.38) demonstrates the effect of the increased roughness in the areas of windblow. An evident amount of blocks stops in this area, collected like in a rack. The defined shape of the red shaped deposit area at the downslope side, results from the densely forested area between the federal road and the area of windblow (Fig. 5.37). The second area of windblow at the middle part of the southern slope shows also an effect on the block run-out (orange shaped area, Fig. 5.38). The rocks descending from the southern part of the

block volume / underground parameters	Min. 0.15x 0.2x 0.25 m	Mean 0.2x 0.25x 0.35 m	Max. 0.5x 0.6x 0.8 m	Blocks 0.8x 1.0x 1.2 m
Transit_1 parameters based on fieldwork; validated ref. to Dorren (2010)				X
Transit_2 increasing roughness (MOH) + 0.1 m at the talus slope + 0.15 m at the valley bottom				
Transit_rough_1 parameters based on fieldwork; validated ref. to Dorren (2010) incorporating areas of windbreakage				X
Transit_rough_2 increasing roughness (MOH) + 0.1 m at the talus slope + 0.15 m at the valley bottom incorporating areas of windbreakage				

Fig. 5.36 Parameter Matrix showing the chosen scenario: the block sizes of the class "Blocks" (0.8 × 1.0 × 1.2 m) were combined with the underground classes "Transit_1" and "Transit_rough_1", where "Transit_rough_1" includes the areas of windblow

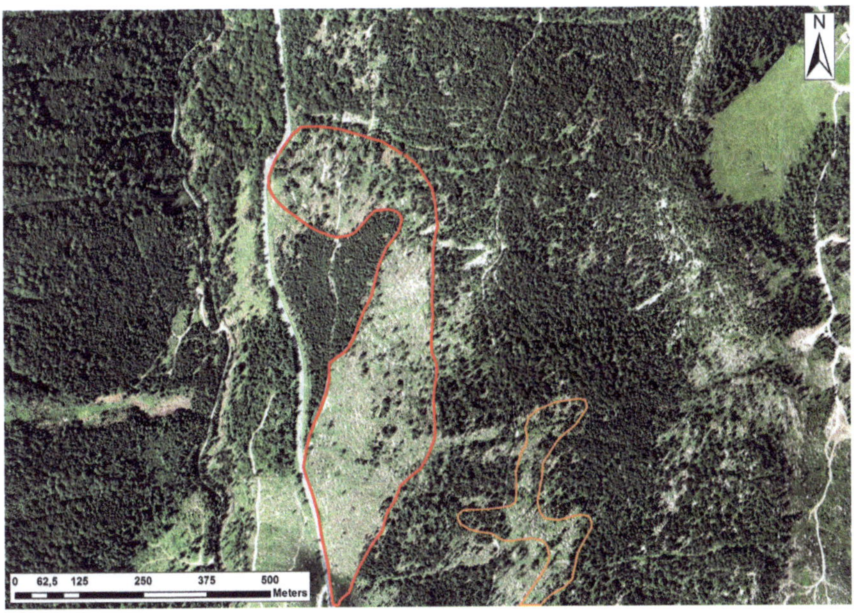

Fig. 5.37 Dominant area of windblow influencing the rockfall run-out at the project site Weißwand. The windblow area is marked with the *red* and *orange shape*. In case of the southern windblow the *orange shape* results from a combination of the forest stand parameters with the underground parameters

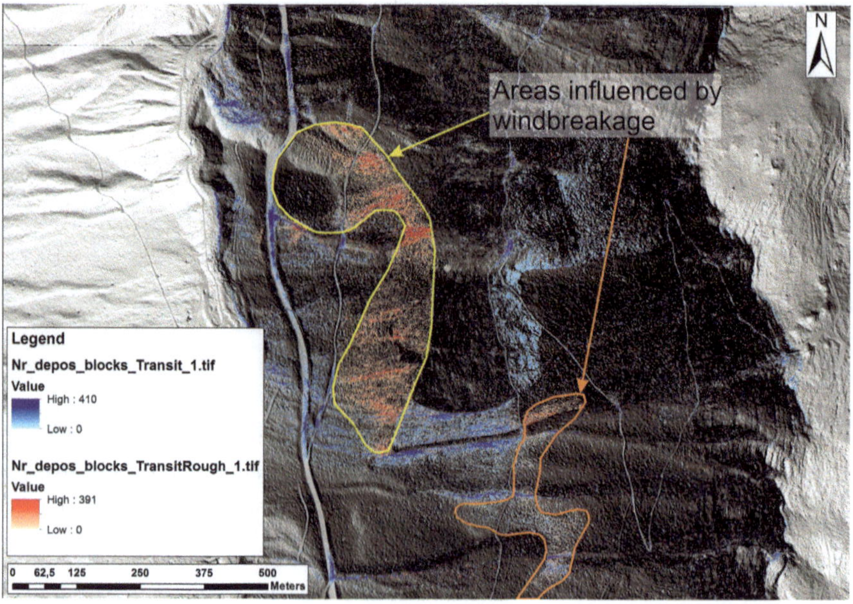

Fig. 5.38 The influence of windblow is visualized by presenting the modelling results of the block class "Blocks" combined with the underground class "Transit_1" and "Transit_rough_1" (where the area of windblow is included). The *yellow* and *orange* shaped area represents the windblow area based on the forest stand and underground parameter dataset

upper rock cliff are collected in this area to a great extent. The absolute maximum of deposited blocks per raster cell for this scenario can be specified as 391. Most raster cells indicate an accumulated block amount of 5 to 30 blocks at the talus slope and 20 to 60 blocks at the federal road or the valley bottom.

5.3.2 RAMMS::Rockfall

Parameter Setting

To compare the results of the parameter study in terms of the hazard situation along the federal road, it was decided to present the following parameters for each scenario:

- Velocity
- Kinetic energy
- Jump height

Each scenario was considered with and without taking the forest stand into account. The presentation of the results will follow the four fragmentation classes ("minimum rockfall", "minimum, mean and maximum fragmentation") of Table 4.6. For each scenario the results of the forested and non-forested sets will be compared. To account for the dispersion range of the descending blocks along the federal road, it was decided to take 40 blocks and 8 shape types into account.

Evaluation of the Field Investigation

Based on the results of field work the homogenous areas for slope and forest stand were determined. Due to the maximum editable underground shapefiles in the first version of RAMMS::Rockfall, the homogenous areas of the underground parameters correspond to the dominating process areas (Fig. 5.39). The forest stand parameters were transferred from GIS and equal the forest-homogenous areas used for Rockyfor3D (Fig. 5.40). The chosen underground parameters are shown in Table 5.18. The forest stand is classified in light, medium and dense forest, which stands for the parameters shown in Table 5.19.

Modelling

Minimum Fragmentation

The scenario “Minimum fragmentation” takes block-volume ranges from approximately 12–33 m^3 (Table 4.6) into account, based on the information obtained from field investigation.

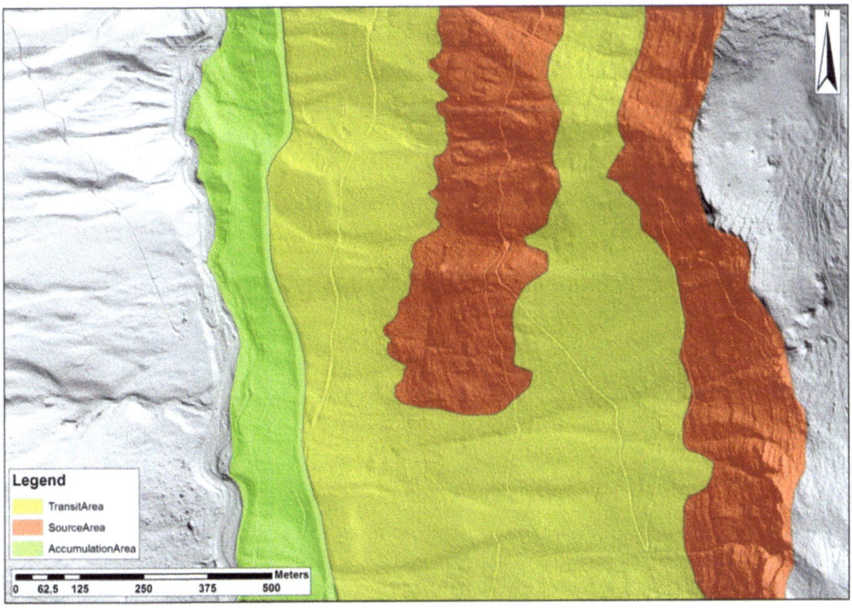

Fig. 5.39 The slope parameters in RAMMS::Rockfall were related to the process areas determined by field investigation and geodata analysis

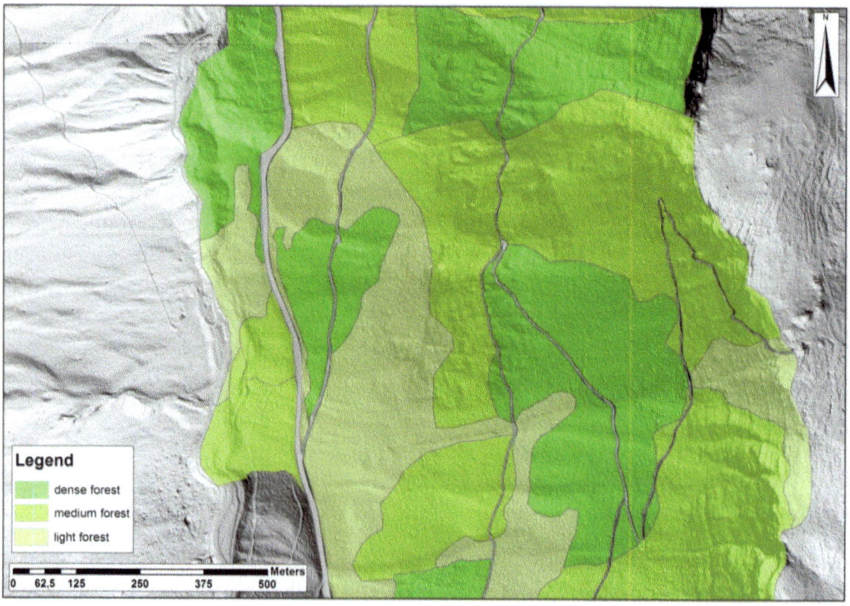

Fig. 5.40 The forest stand was categorized according to the forest classes provided by the code RAMMS::Rockfall based on the homogenous forest areas determined by field work and aerophoto analysis

Table 5.18 Visualisation of the Parameters selected for the slope at the Weißwand Project site

Process area	Parameter set	Mu_Min.	Mu_Max.	Kappa	Epsilon	Beta	Ground drag
Source area	Hard	0.55	2.00	3.00	0.00	100.0	0.4
Transit area	Medium hard	0.40	2.00	2.50	0.00	125.0	0.5
Accumulation area	Medium	0.35	2.00	2.00	0.00	150.0	0.6

The source area was characterized by a “hard” terrain, the talus slope by “medium hard” terrain and the accumulation area by “Medium” terrain

Table 5.19 Forest stand parameters chosen for the homogenous areas, classified according to the forest density

Forest density	Forest height	Drag force
Light forest	5.00	1000
Medium forest	5.00	1500
Dense forest	5.00	2000

Figure 5.41 shows the velocity distribution of the falling blocks, which suggests a minor dependence of the forest stand. The modelling results for the non-forested slope are presented in the left part of Fig. 5.41; the velocities for the forested slope

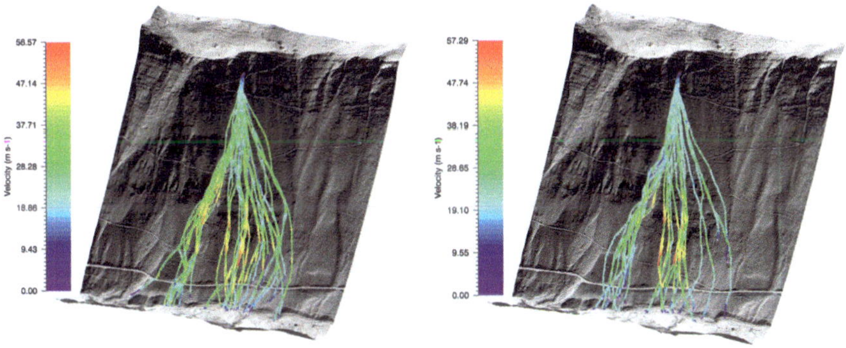

Fig. 5.41 Velocity distribution and block run-out for the scenario "Minimum fragmentation", implying a low fragmentation rate of the critical block with a volume range of 12–33 m^3. *Left* The block run-out and velocities without a consideration of the forest stand. *Right* The scenario including the forest stand

are shown in the right part of Fig. 5.41. The maximum speed of 40–56 m/s occurs below the steep rock cliff in the middle part of the slope. Towards the federal road, the mean velocity decreases to 20–30 m/s.

Due to the high rock volume the kinetic energies show very high results (Fig. 5.42). The highest kinetic energies with values between about 100,000 and 130,000 kJ, also occur at the bottom of the rock cliff at the middle part of the slope. The maximum kinetic energies can be specified with 152,186 kJ for the non-forested scenario and with 140,982 kJ for the forested slope. The kinetic energies along the federal road are suggested to be in average about 40,000 kJ for the non-forested slope and about 30,000 kJ the forested slope.

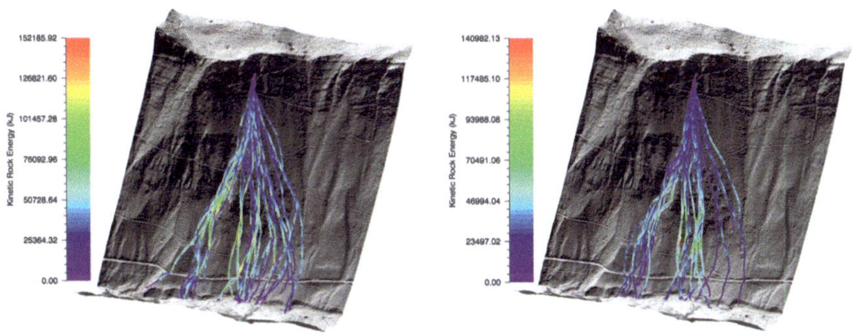

Fig. 5.42 Kinetic energy distribution and block run-out for the scenario "Minimum fragmentation", implying a low fragmentation rate of the critical block with a volume range of 12–33 m^3. *Left* The block run-out and kinetic energies without a consideration of the forest stand. *Right* The scenario including the forest stand

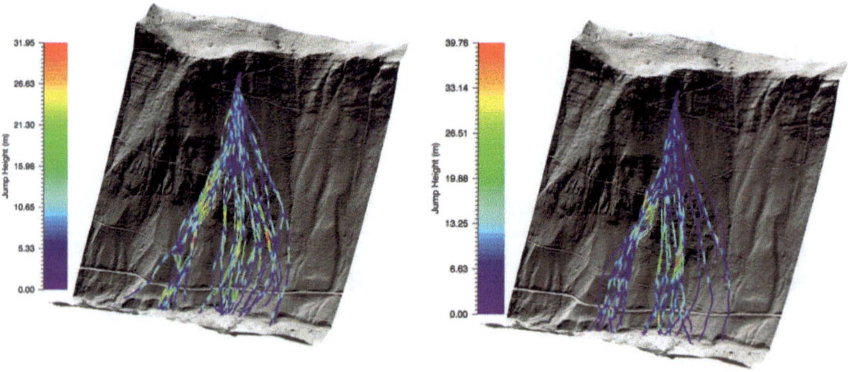

Fig. 5.43 Jump height distribution and block run-out for the scenario "Minimum fragmentation", implying a low fragmentation rate of the critical block with a range of 12–33 m^3. *Left* The block run-out and jump heights without a consideration of the forest stand. *Right* The scenario including the forest stand

The jumping height represents the distance of the centre of mass in relation to the slope surface in vertical direction (Fig. 5.43). Due to this fact the highest jumping heights of 30–40 m (in average for the forested and non-forested slope) occur at the bottom of the rock cliffs at the middle part of the slope. The maximum jumping height for the non-forested scenario can be named with 31.95 m, for the forested slope with 39.76 m. Along the federal road, the maximum jumping height decreases to about 10 m in average for the forested and non-forested slope.

For the "Minimum fragmentation" scenario the blocks all descend to the Schwarzbach at the valley bottom due to the high rock volume. The dispersion of the trajectories at the federal road would measure approximately 600 m in width.

Mean Fragmentation

The scenario "mean fragmentation" takes block-volume ranges from approximately 1.5–4.2 m^3 (Table 4.6) into account. Figure 5.44 shows the velocity distributions without taking the forest stand into account (left) and under consideration of the forest stand (right). The velocity distribution suggests an increasing dependence on the forest stand. The maximum speed of 50–62 m/s occurs below the steep rock cliff in the middle part of the non-forested slope. On the forested slope the maximum velocities decrease to a maximum of 20–30 m/s. Towards the federal road the speed averages between 20 and 50 m/s for the non-forested and 5–15 m/s for the forested slope.

The highest kinetic energies occur at the below the steep rockcliffs at the middle part of the slope, which is accordance with the results from the velocity distribution. The maximum kinetic energies can be specified as 15,000–19,000 kJ for the non-forested slope (Fig. 5.45, left) and as 4000–6000 kJ for the forested slope (Fig. 5.45, right). The kinetic energies along the federal road are suggested to vary in average between 3000 and 9000 kJ for the non-forested slope and about 500–2000 kJ for the blocks reaching the road at the forested slope.

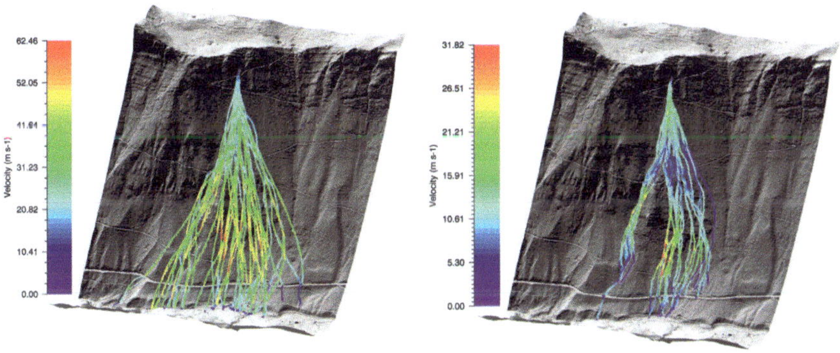

Fig. 5.44 Velocity distribution and block run-out for the scenario "mean fragmentation", implying a mean fragmentation rate of the critical block with a volume range of 1.5–4.2 m^3. *Left* The block run-out and velocities without a consideration of the forest stand. *Right* The scenario including the forest stand

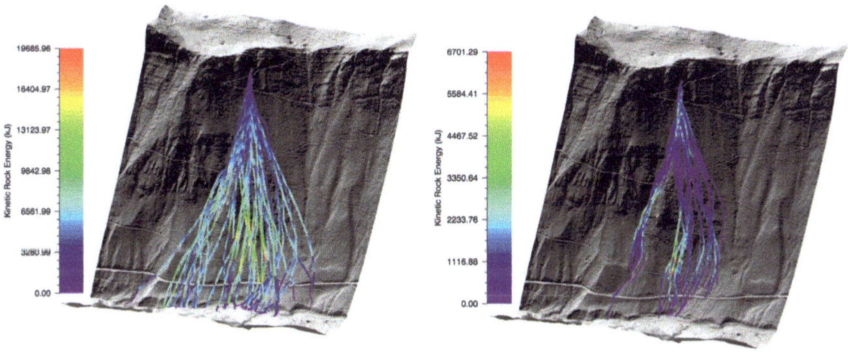

Fig. 5.45 Kinetic energy distribution and block run-out for the scenario "mean fragmentation", implying a mean fragmentation rate of the critical block with a volume range of 1.5–4.2 m^3. *Left* The block run-out and kinetic energies without a consideration of the forest stand. *Right* The scenario including the forest stand

The jumping height represents the distance of the centre of mass in relation to the slope surface in vertical direction. Due to this fact the highest jumping heights of about 40 m at the non-forested slope occur at the bottom of the rock cliffs at the middle part of the slope, whereas the maximum jumping height for the forested scenario can be named with 11 m (Fig. 5.46). In average the jumping heights do not exceed 1–2 m in average for the forested slope. Along the federal road, the maximum jumping height decreases to about 5–10 m for the non-forested and to 1 m for the forested slope.

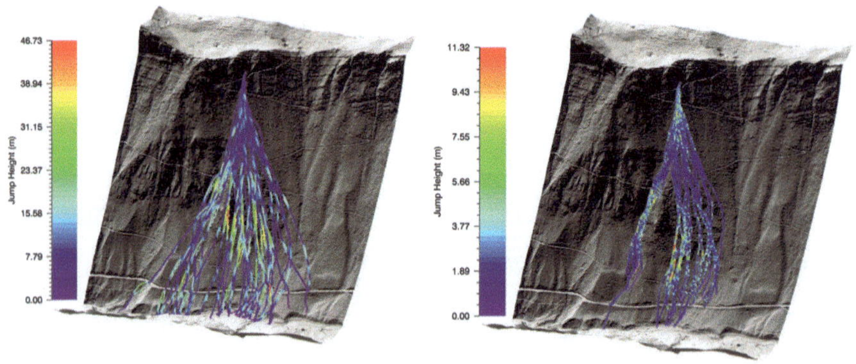

Fig. 5.46 Jump height distribution and block run-out for the scenario "mean fragmentation", implying a mean fragmentation rate of the critical block with a volume range of 1.5–4.2 m^3. *Left* The block run-out and jump heights without a consideration of the forest stand. *Right* The scenario including the forest stand

The results of the "Mean fragmentation" scenario (where the dimensions of the block axes are halved compared to the ones of "Minimum fragmentation") show an increasing dependence on the forest stand density.

Maximum Fragmentation

The scenario "maximum fragmentation" takes block-volume ranges from approximately 0.2–0.7 m^3 (Table 4.6) into account. Figure 5.47 shows the results obtained for modelling without taking the forest stand into account (left) and under consideration of the forest stand (right). The velocity distribution of the falling blocks is illustrated at the top row and suggests an evident dependence on the forest stand. The maximum speed of 50–67 m/s occurs below the steep rock cliff in the middle

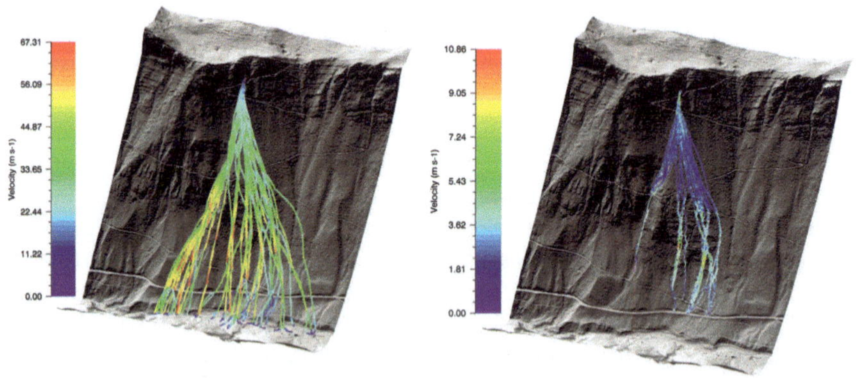

Fig. 5.47 Velocity distribution and block run-out for the scenario "maximum fragmentation", implying a high fragmentation rate of the critical block. *Left* The block run-out without a consideration of the forest stand. *Right* The scenario with consideration of the forest stand

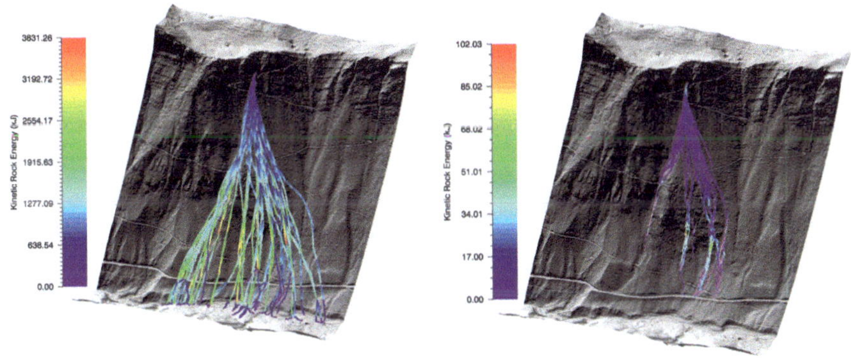

Fig. 5.48 Kinetic energy distribution and block run-out for the scenario "maximum fragmentation", implying a high fragmentation rate of the critical block. *Left* The block run-out and kinetic energies without a consideration of the forest stand. *Right* The scenario with consideration of the forest stand

part of the non-forested slope. On the forested slope the maximum velocities decrease to a maximum of 8–10 m/s. Towards the federal road the speed averages between 20 and 45 m/s for the non-forested and 1–5 m/s for the forested slope. The highest kinetic energies occur at the below the steep rockcliffs at the middle part of the slope, which is accordance with the results from the velocity distribution (Fig. 5.47).

The maximum kinetic energies can be specified as 3000–3800 kJ for the non-forested slope and as 80–100 kJ for the forested slope (Fig. 5.48). The kinetic energies along the federal road are suggested to be in average about 1000–1500 kJ for the non-forested slope and about 15–20 kJ for the blocks reaching the road at the forested slope.

The jumping height represents the distance of the centre of mass in relation to the slope surface in vertical direction. Due to this fact the highest jumping heights of about 40 m (in average for non-forested slope) occur at the bottom of the rock cliffs at the middle part of the slope (Fig. 5.49). The maximum jumping height for the forested scenario can be named with 2.3 m. In average the jumping heights do not exceed 0.8 m in average for the forested slope. Along the federal road, the maximum jumping height decreases to about 10 m for the non-forested and to 0.7 m for the forested slope.

The results of the maximum fragmentation scenario (where the dimensions of the block axes are quartered compared to the ones of "minimum fragmentation") show an evident dependence on the forest stand density.

Minimum Rockfall

The scenario "minimum rockfall" takes block-volume ranges from approximately 0.2–0.7 m^3 (Table 4.6) into account. Figure 5.50 shows the results of the rock velocities for modelling without taking the forest stand into account (left) and under consideration of the forest stand (right). The velocity distribution suggests an

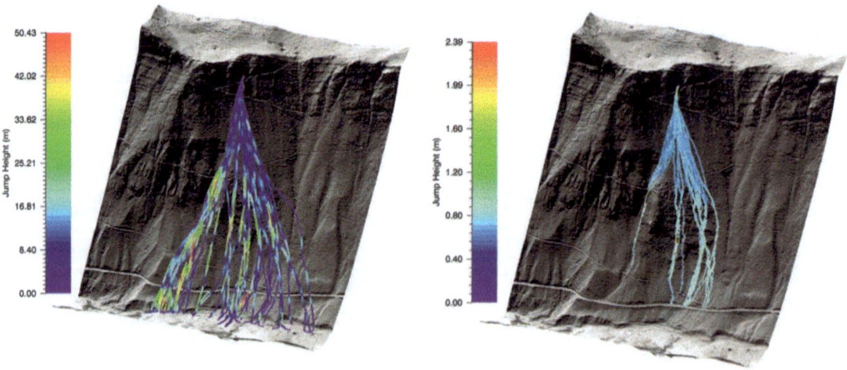

Fig. 5.49 Jump height distribution and block run-out for the scenario "maximum fragmentation", implying a high fragmentation rate of the critical block. *Left* The block run-out and jump heights without a consideration of the forest stand. *Right* The scenario including the forest stand

Fig. 5.50 Velocity distribution and block run-out for the minimum volume class "minimum rockfall", implying the highest fragmentation rate taking a volume range of 0.1–1 m^3 into account. *Left* The velocities and rock run-outs without accounting for the forest stand. *Right* The velocities and trajectories with consideration of the forest stand

evident dependence on the forest stand (Fig. 5.50). The maximum speed of 50–72 m/s (for the non-forested slope) occurs below the steep rock cliff in the middle part slope. On the forested slope the maximum velocities decrease to a maximum of 3–4 m/s. Towards the federal road the speed averages between 40 and 60 m/s for the non-forested slope. Taking the forest stand into account, the rocks will not run-out to the federal road at this volume class.

Comparable to the other scenarios the highest kinetic energies between 800 and 1200 kJ occur at the talus slope below the rockcliffs at the middle part of the slope. For the forested slope the kinetic energies show drastically decreased kinetic energies, which vary between 0.5 and 1.0 kJ (Fig. 5.51). The kinetic energies along

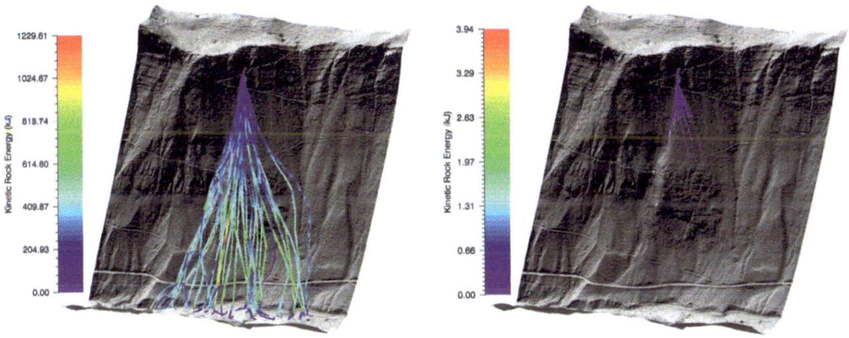

Fig. 5.51 Kinetic energy distribution and block run-out for the minimum volume class "minimum rockfall", implying the highest fragmentation rate taking a volume range of 0.1–1 m^3 into account. *Left* The kinetic energies and rock run-outs without accounting for the forest stand. *Right* The kinetic energies and trajectories with consideration of the forest stand

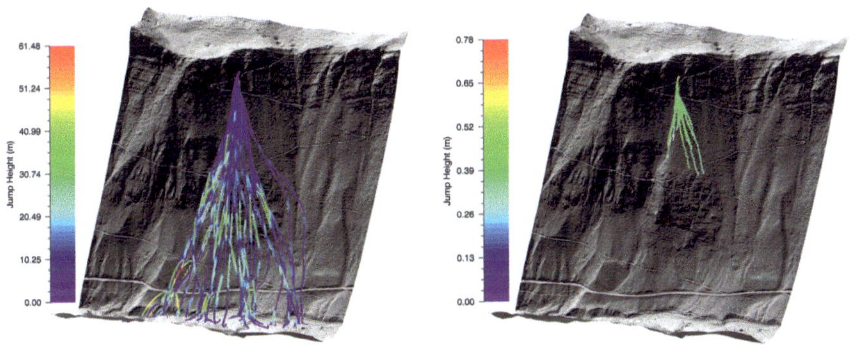

Fig. 5.52 Jump height distribution and block run-out for the minimum volume class "minimum rockfall", implying the highest fragmentation rate taking a volume range of 0.1–1 m^3 into account. *Left* The jump heights and rock run-outs without accounting for the forest stand. *Right* The jump heights and trajectories with consideration of the forest stand

the federal road are suggested to be in average about 500–1000 kJ for the non-forested slope.

The jumping height represents the distance of the centre of mass related to the slope surface in vertical direction. Thus the highest jumping heights of about 40 m (in average for non-forested slope) occur at the bottom of the rock cliffs at the middle part of the slope (Fig. 5.52). One trajectory suggests an increased jumping height between the cliffs and the federal road of 60 m. The maximum jumping height for the forested scenario can be named with 0.5–0.6 m. In average the jumping heights do not exceed 0.8 m in average for the forested slope. Along the federal road, the maximum jumping height decreases to about 5–20 m for the non-forested slope.

The results of the "Minimum rockfall" scenario (where the dimensions of the block axes equal the minimum editable block volumes in RAMMS::Rockfall) suggest a high dependence on the forest stand density. The max dispersion cone measures approximately 580 m in width along the federal road for the non-forested slope. The run-out of the rocks for the forested slope end in the upper part of the talus slope, above the rock-cliffs at the middle part of the slope.

References

Barton N, Choubey V (1977) The shear strength of rock joints in theory and practice. Rock Mech 10:1–54

Bayerisches Landesamt für Umwelt (LfU) [Hrsg.] (2014) Gefahrenhinweiskarte Alpen mit Alpenvorland - Landkreis Berchtesgadener Land. – 78 S., München (Bayrisches L.-Amt für Umwelt, pdf-version)

DGGT—Deutsche Gesellschaft für Geotechnik e.V. (2004) Einaxiale Druckversuche an zylindrischen Gesteinsprüfkörpern. Neufassung der Empfehlung Nr. 1 des Arbeitskreises "Versuchstechnk Fels" der Deutschen Gesellschaft für Geotechnik e.V.- Bautechnik, 81, 825–834

Dorren LKA (2012) Rocky for 3D (v5.0) revealed—Transparent description of the complete 3D rockfall model. ecorisQ paper (www.ecorisq.org) 31 p

ISRM—International Society for Rock Mechanics (1978) Suggested methods for the quantitative description of discontinuities in rock masses. Commission on Standardization of Laboratory and Field Tests. Int J Rock Mech Min Sci Geomech Abstr 15(4):319–368

Priest SD (1993) Discontinuity analysis for rock engineering. Chapman & Hall, London, 473 p

Thuro K (1996) Bohrbarkeit beim konventionellen Sprengvortrieb.- 145 S., Münchner Geol. Hefte, Reihe B: Angewandte Geologie: Heft 1 (elektronische Publikation)

Chapter 6
Discussion

6.1 Case Study: Potential Planar Rock-Slide at the Wachterl-Horn

6.1.1 Mapping of the Critical Block Subjected to Planar Failure

The mapping of the block at the "Wachterl Steig" was accomplished using mainly a measuring tape and a folding rule, due to the steep terrain around the block. The hard accessibility and visibility prohibited the possibility of using photogrammetric or laserscanning tools.

Due to this fact the dimensions of the block and the failure cave/surface have been mapped using a folding rule and a measuring tape. The uncertainty in measuring the dimensions for the mapping underneath the block could be specified as approximately 5–10 cm referring to the cave entrance. The main uncertainty in estimating the volume of the critical block is the width along the slope. There is an evident discrepancy between the outer measured width of the block based on the morphological character and the width of the failure surface underneath the block. The "outer" width can be specified as 4.70 m whereas the width of the failure cave measures at least 8.50 m at the accessible area, which suggests that the morphological block does not correspond to the critical mass in total. Since the results of detailed field work show, that the failure surface outcrops at the south side of the block, it is suggested that the critical mass is larger than the morphological block (Fig. 5.1).

The assessment of joint persistence through the block is based on the joint traces at the rock surface and the joint orientation. As far as possible the traces were mapped from the inner and outer side at the block. For the study of a potential fragmentation scenario (Fig. 5.1) only the main joints were taken into account. Nevertheless this is an assumption, since it is not possible to follow the joints through the rock mass.

B. Sellmeier, *Quantitative Parameterization and 3D-run-out Modelling of Rockfalls at Steep Limestone Cliffs in the Bavarian Alps*, Springer Theses,
DOI 10.1007/978-3-319-24510-2_6

6.1.2 *Recording of the Shear Parameters and Uniaxial Compressive Strength Testing*

The quantitative recording of the shear parameters included the determination of the roughness coefficient (JRC), the joint compressive strength (JCS) and the validation of the JCS by performing uniaxial compressive strength (UCS) tests in the laboratory.

The JCS tests were performed as 200 clustered tests at the failure surface underneath the block. Each test series included 10 single tests at a sample area of 10 by 10 cm. The test number correlates with the one suggested by Barton and Choubey (1977), whereas we decided to reduce the test area from a square meter to 10 by 10 cm due to a lack of space at the failure cave underneath the block (Barton and Choubey 1977: 11). The test areas were equally distributed over the failure surface including the two main roughness areas (ISRM 1978). The sample areas were chosen very carefully in terms of the following influence factors (Goudie 2006; Barton and Choubey 1977):

- Structure of the limestone, meaning nappes or flakes
- Surface roughness, since the hammer head has got a convex shape and irregularities may be crushed before the plunger tip reaches the rock
- Bio- or Sintercrusts

Due to irregular moisture content the Schmidt-Hammer results could be also influenced, but this uncertainty could not have been eliminated in the current test settings (Goudie 2006). Even though the named influencing factors were considered during test performing, the results show an evident scatter.

Figure 6.1 visualizes the scatter of every test series in a box-whisker plot, where the labels of the test locations are plotted versus the JCS in MPa. With an increasing labeling number, the distance from the cave entrance increases, where Sh1 is 0.5 m away from the cave entrance and Sh20 is 6.5 m away from the cave entrance. The results show an evident wide scatter, which cannot be correlated with the distance from the cave entrance. Even though the test locations were chosen in consideration of the above mentioned factors of influence, the results show a dispersion range of up to 85 MPa (from a Min of 25 MPa to a Max. of 110 MPa) at a test area of 10 by 10 cm. The partly high scatter of the test results is suggested to be explained by two dominant influencing factors: Firstly the formation of little cracks causing thin nappes at the rock surface and secondly the low scale surface roughness. Due to the rebound of the Schmidt Hammer little nappes are crushed before the rock mass strength is tested. Performing acoustic tests by knocking at the rock surface in advance can localize hallow nappes, but not narrow opened cracks. The effect of small scale surface roughness is related to the proportion of the hammer head and the surface roughness. If the roughness scale is similar or lower than the diameter of the hammer pile, the convex hammer pile can be set on small roughness peaks and be shifted or moved during testing. This can lead to a different testing angle in

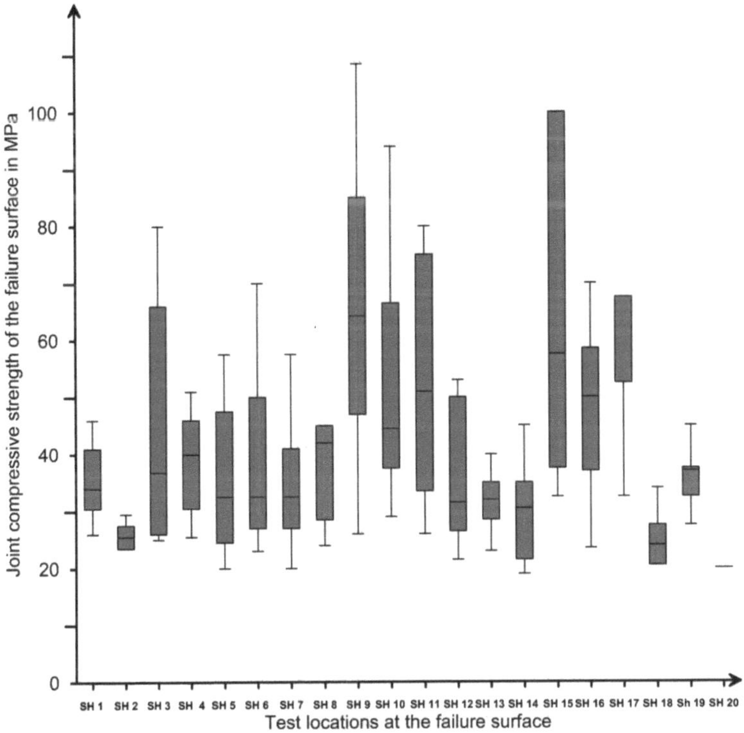

Fig. 6.1 Joint compressive strength (JCS) plotted against the fracture roughness (ISRM 1978), including the rock bridges as an extra class. The mean values of the JCS are increasing with decreasing roughness. The JCS (23–38 MPa) of the rock bridges is significantly low

relation to the rock surface which might result in different values even on small sample areas.

The mean values of the test series are shown in Fig. 6.2 where the red dots represent the arithmetic JCS-mean values of ten single tests each and the blue triangles stand for the results corrected according to Barton and Choubey (1977) (meaning the mean of the best five values). The compressive strength is plotted versus the distance from the cave entrance. The mean values of the 10 single tests vary between 25 and 73 MPa, whereas the corrected mean values vary between 26 and 92 MPa. For the stability analysis of the current case it is meaningful to consider the mean values of the 10 single tests, since the suggested correction implies the elimination for the 5 lowermost JCS values of every test series (Barton and Choubey 1977), which indicates an overestimation of the jontwall compressive strength in this context.

The results of the uniaxial compressive strength tests vary between 80 and 115 MPa (Fig. 6.3). This could be classified as high to very high (ISRM 1978). According to this classification the limestone is supposed to have a uniaxial

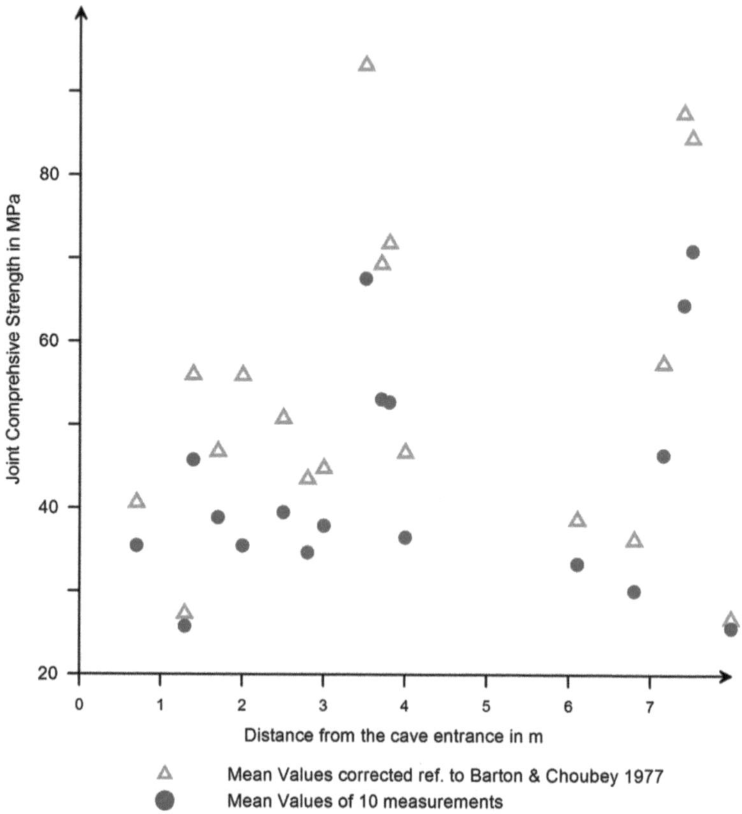

Fig. 6.2 Plot of the joint compressive strength (JCS) and the distance from the cave entrance. The *red dots* show the mean value out of 10 single measurements. The blue triangles represent the mean values of ten single tests corrected according to Barton and Choubey (1977)

compressive strength of about 150 MPa, whereas Sandstones would have a UCS Range of about 50–100 MPa (Fig. 4.4; ISRM 1978). Thus the UCS of the tested dolomized limestone is set to be in the strength field in between of sandstone and limestone. The tested limestone of the Dachstein-Formation contains an increased dolomite content, which leads to a narrow discontinuity spacing. The samples with dimensions of 10 cm in length and 5 cm in width are drawn through by numerous joints. The photo material and sample analysis after the test shows a mode of failure often related to one of the described discontinuities leading to a reduced material strength. A considerable portion of joints contain calcite fillings, which could also lead to a reduction in strength.

Summing up the general mean of the Schmidt-Hammer tests (Fig. 6.2) varies between 30 and 40 MPa, whereas the uniaxial compressive strength tests show values between 82 and 115 MPa. The scatter of the maximum values of the Schmidt

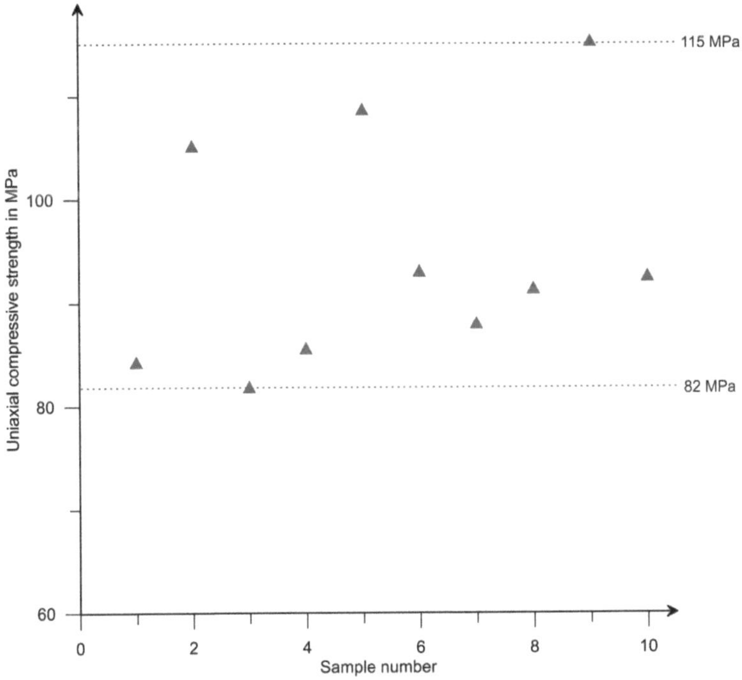

Fig. 6.3 Diagram of the 10 uniaxial compressive strength tests. The 10 samples show a compressive strength between 82 and 115 MPa. The grey dashed lines indicate the minimum and the maximum values

Hammer Tests (SH 9 and SH16) includes the range of results of the UCS tests. Nevertheless the mean of the joint compressive strength is suggested to be about 50 % lower than the uniaxial compressive strength. This reduction of 50 % is even more than the suggested strength reduction about 25 % for weathered joints compared to UCS values of fresh material (Barton 1973: 328).

The joint roughness coefficient (JRC) was recorded along three cross sections in dip direction of the failure surface underneath the block. The way of evaluation is in detail described in Sect. 4.1.2, p. 25. The recording was performed using a fault gauge of 260 mm in length. Even though the start and end locations of the fault gauge were marked as possible, narrow overlapping areas of approximately 1 cm could not be excluded.

The interval of JRC recording was varied in steps from 5, 10 to 20 mm (McCarroll and Nesje 1996; McCarroll 1997). The results suggest an evident dependence of the joint roughness coefficient on the interval length of the evaluation, indicating a decrease of the JRC with increasing length of recording-intervals. For the current study a block with a failure surface of a length of at least 9.0 m is considered. Thus even the maximum selected evaluation interval of 20 mm is suggested to be very narrow-spaced compared to the relevant length of the failure surface. Thus the JRC results of this study have to be interpreted in relation to the length of the failure surface (Barton and Choubey 1977: 40 ff.). The evaluated results represent a fracture roughness of the failure surface at low scale meaning in a high resolution. It is to note that the undulation or unevenness of this failure surface was not recorded in a quantitative way (Fecker 1978: 116 ff.), but only in a qualitative one referring to ISRM (1978), indicating that the undulation of the failure surface is suggested to be "undulating, rough" respectively "undulating, smooth to rough" (Sect. 5.1.1; Fig. 5.2). It is suggested that the larger scale roughness and the undulation would have an increased effect on the stability of the block compared to the small scale roughness due to the following arguments:

- Small scale roughness contacts (implying amplitudes of several cm over a distance of 1–5 cm) are suggested to be smoothened or sheared off due to the high normal load provided by the critical block (Sect. 5.1.5).
- The main area of contact is suggested to be dominated by undulation contacts of larger wavelength (suggesting an undulation of about 10–15 cm in amplitude over a distance of 30–40 cm)

The first argument is based on a study of Krahn and Morgenstern suggesting that roughness peaks will be smoothened during the shearing process (Krahn and Morgenstern 1979: 131). Figure 6.4 illustrates the suggested relation between small scale roughness and undulation for the current study case.

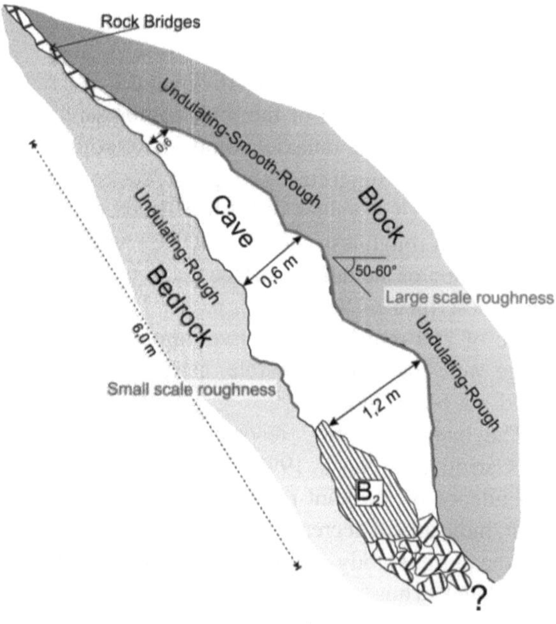

Fig. 6.4 Cross section through the failure "cave" illustrating the suggested relation between small scale roughness (recorded in terms of the JRC) and large scale roughness (undulation). The undulation was only qualitatively recorded for the current study case

6.1.3 Determination of the Density

The results of the density determination (Table 5.5, p. 50) suggest that the dry bulk density of the carbonates can be considered nearly the same as the calculated density. This implies that the samples have absorbed no water to the lack of pore space.

6.1.4 Limit Equilibrium Analysis

The uncertainties in terms of measuring the block dimensions are described in Sect. 6.1.1. The limit equilibrium analysis was accomplished for two assumed rock volume scenarios varying the width of the failure surface across the slope. The variation in rock volume shows a direct influence on the driving forces and herewith on the shear stress, due the direct dependence on the force of gravity. Since quotient of the factor of safety (FS) is independent of the force of gravity, the volume of the mass of failure has no influence on the FS:

$$\vartheta = \frac{\tan \varphi}{\tan \alpha} \tag{6.1}$$

where υ is the factor of safety, α is the slope inclination angle and φ is the friction angle.

The result of Sect. 5.1.5 show, that for the current study case (without taking the cohesion into account), the factor of safety would be 0.5—instable. To achieve a labile equilibrium state cohesion of 37 kN/m^2 would be necessary, which is suggested to be a realistic value range for hard rocks.

To assess the influence of the surface roughness "i" on the stability stage, we back calculated the angle which would be required to compensate the discrepancy between the basic friction angle and the slope angle, which would be 20° for a measured slope angle of 55° and an assumed basic friction angle of 35° (Cruden and Hu 1988; Hoek et al. 1998). The influence of the JRC-range and the measured JCS values (of the current study) on the dilation angle "i" are visualized in Figs. 6.5 and 6.6. The influence of the JCS is suggested to decrease with decreasing JRC. To compensate the 20° discrepancy between slope and basic friction angle an additional roughness namely a JRC of 7 for a mean JCS of 40 MPa would be required (Figs. 6.5 and 6.6).

In the current study the required dilation angle "i" and the necessary cohesion were characterized independently from each other. In practices the resisting forces will be enhanced by a combination of cohesion and surface roughness. In the field of hard rock mechanics the transition from cohesion to roughness contacts is not clearly definable, due to a gradually transition from rock bridges to roughness contacts. For the current study in the dolomized limestone of the Dachstein-Formation a

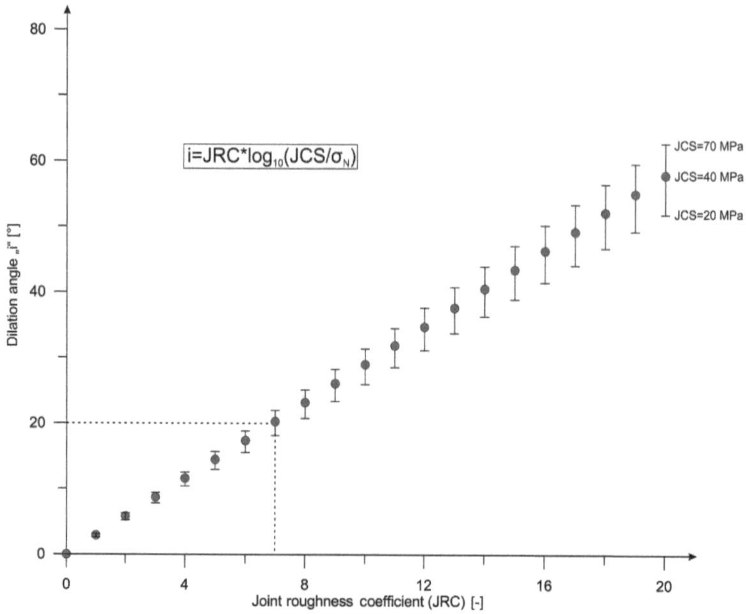

Fig. 6.5 Shows the dilation angle "i" plotted for the three measured JCS values: 40 MPa (mean), 20 MPa (general minimum) and 70 MPa (general Maximum) versus the Joint roughness coefficient (JRC) for scenario 01

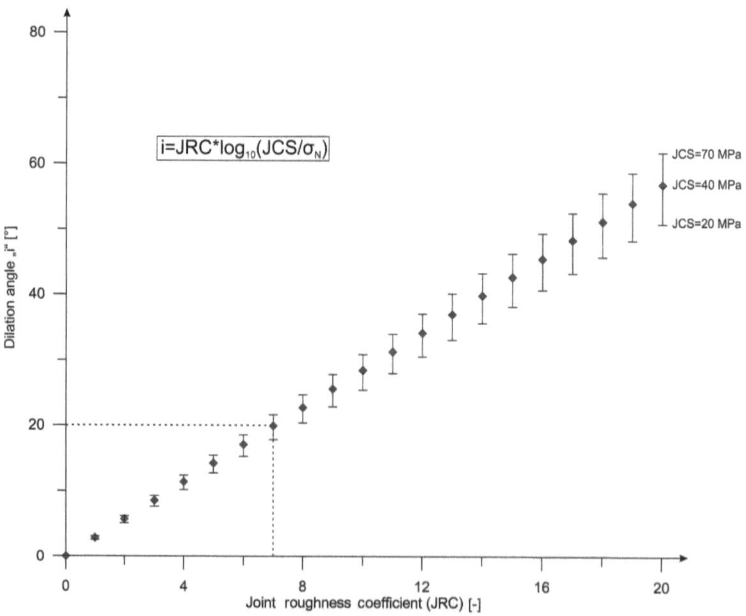

Fig. 6.6 Shows the dilation angle "i" plotted for the three measured JCS values: 40 MPa (mean), 20 MPa (general minimum) and 70 MPa (general Maximum) versus the Joint roughness coefficient (JRC) for scenario 02

scheme sketch for the failure of rock bridges was developed based on field observations in the failure surface underneath the critical block (Fig. 6.7). In a first stage cracks form along a potential shear plane. Due to increased aligned stresses the cracks start growing together, the opening width enhances. As the crack density increases the loosened and sheared material can be washed out and in addition the potential failure plane can be enlarged due to karst processes. The failure surface is composed of rock bridges (cohesive parts) and partly interlocking roughness contacts (roughness and friction parts). Interlocking roughness contacts can lead to locally increased shear stresses, potentially being released during the shearing failure of the asperities. Due to gradient failure of rock bridges, the shear stresses are suggested to be distributed to the remaining rock bridges and interlocking asperity contacts. These transition stages from rockbridges to roughness contacts suggest a parallelization between the diagrams of Figs. 6.5 and 6.6 and the sketch of Fig. 6.7. A range of medium to high dilation angles (30°–60°, JRC = 11–20) is suggested to

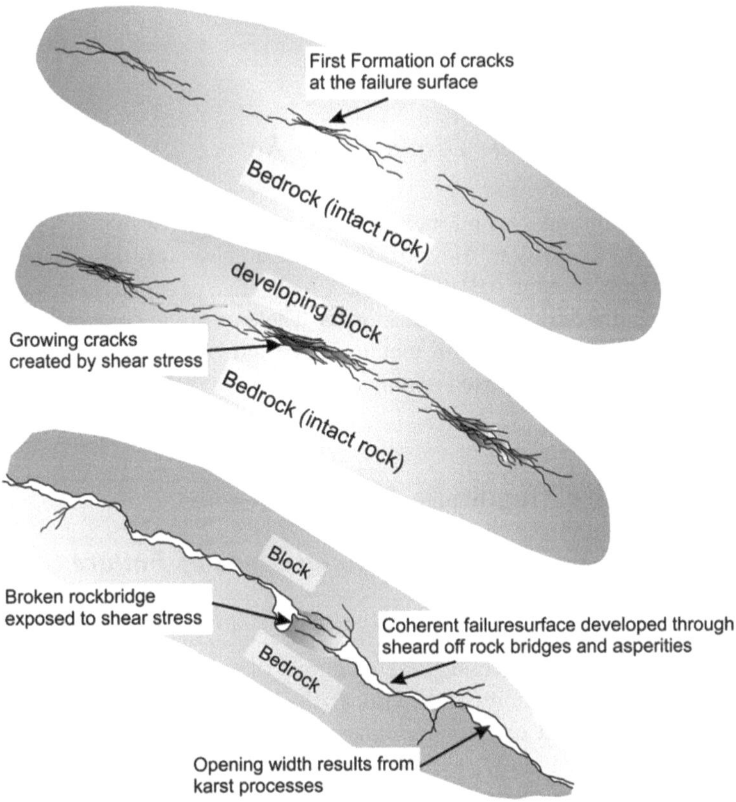

Fig. 6.7 Sketch illustrating the development of rock bridges in the carbonate rocks of the Dachstein-Formation: in the initial stage cracks are formed in a potential shear plane. Since the shear-or driving force increases, the cracks start growing together. Once the cracks get connected; the influence of water and karst processes enlarges the shear plane to an opened failure plane

correspond to dissolving rockbridges or strong interlocking roughness contacts (for example the broken rockbridge of Fig. 6.7). A medium dilation angle between 15° and 30° (JRC = 5–11)is suggested to interlocking, rough undulation contacts, like in the left part of the bottom cross section of Fig. 6.7. A dilation angle between 0° and 15° (JRC = 0–5) is suggested to cover the general range of surface roughness contacts along a failure surface. The suggested dilation angle/JRC ranges are based on the recorded mean joint compressive strength values during field investigation. Therefor the suggested model refers to the current study in carbonate rocks of the Dachstein-Formation.

For the current study the influence of joint water pressure was neglected in the limit equilibrium analysis. The high opening width of the failure surface, the tension crack at the top of the block and the free discharge at the toe of the block provide certain evidence that precipitation water enters the system at the tension crack at the top of the failure surface, but also drains off at the toe of the block.

During the whole observation period (even after heavy rain fall or during snow melt), the failure surface provided more or less wet, but never water filled conditions.

6.1.5 Monitoring of the Displacement Rates

The installed monitoring system, including 4 strain gauges, was installed for one season from October 2012 to May 2013. The data show very low displacement rates of hundredth of mm (0.0–0.7 mm), which are suggested to results from temperature changes during the winter months. The data are to be considered as an indicating value. The time frame of monitoring period is too low for extrapolating the data over an extending period of time.

6.2 Quantitative Magnitude Assessment

6.2.1 Scanline Analysis and Potential Mode of Failure

To assess the dominant discontinuity sets a "virtual" scanline analysis was performed along a forest road, crossing the main part of the source area at the project site (Sect. 4.2.1). In the vicinity of the critical block two scanlines were recorded to obtain basic and detailed data for the kinematic analysis of the block and information about the discontinuity spacing.

The discontinuity sets were determined due to a contour plot, evaluated by hand. The manual determination of the joint sets by hand includes a certain proportion of subjectivity. The selection of joint-sets-window-outlines was accomplished in combination of the joint data and the knowledge from field work. In the following paragraph the results of the three scanlines recorded at the source area are discussed: For joint set K2 the results vary between 2 similar oriented joint sets ("virtual" scanline and scanline 02) and a grouped joint set K2 for scanline 01. Joint set K2 belongs to the steep north-west dipping discontinuity set, where the maximum scatter in orientation can be specified with 14° (75°–89°) for the dip angle and 55° in terms of the dip direction (299°–354°). Nevertheless we suggest clustering the joint sets to one set since the effect of forming the third joint set in terms of an orthogonal discontinuity system is given despite of the scatter. The joint set K1 represents the discontinuity set dipping almost parallel to the slope orientation. The variation in dip angle can be specified with 6° (53°–59°) and the scatter in terms of dip direction is 22° (229°–251°). The results of the scanline-path are suggested to provide the most reliable results being in accordance with the observations from fieldwork with an orientation of 251/53. The bedding orientation shows the lowermost variation in total with 9° (39°–48°) in dip and 14° (69°–83°) in dip direction. Due to the bedding being the primary and sedimentary caused discontinuity set this result is consequential.

The two scanlines at the vicinity of the block were recorded referring to the suggested criterions of Priest and Hudson (1981) and Priest (1993). Due to the stepped, partly vegetated terrain at the carbonate rock faces we recorded two horizontal scanlines, but no vertical ones. Thus the recorded data represent exceptionally the results of discontinuity evaluation in horizontal direction. Each scanline measured 20 m due to the limited access around the critical block. The consequence is a restricted amount of discontinuity data directly recorded via scanline analysis, where scanline 01 contains 35 entries and scanline 02 contains 21 entries. For evaluating the joint spacing along the scanline, the dependence of the joint data on the scanline orientation is important. The scanline 02 had to be divided into two 10 m sections providing in average 10 entries per section. Due to the low amount of discontinuity data a statistical analysis was not carried out for this scanline. For scanline 01 (with 35 entries) the joint spacing referring to the scanline was analyzed (Table 6.1). The analysis includes the minimum, mean and maximum values for the joint spacing for each suggested discontinuity set.

As described due to the low amount of data, the entries considered the bedding can be specified with 6 joints, the entries of joint set K1 with 5 joints and the entries of joint set K2 with 15 entries (which are the discontinuities belonging to determined joint sets in DIPS Rocscience). For the bedding a mean joint spacing of 1.39 m which could be specified as "wide" referring to ISRM (1978: 334). For joint set K1 the spacing is about 0.9 m and for joint set K2 about 0.6 m, which could be

Table 6.1 Evaluation of the joint spacing along scanline 01 at the vicinity of the critical block

Joint set	Joint spacing [m]		
	Minimum	Mean	Maximum
Bedding (6 Joints)	0.92	1.39	1.87
Joint set K1 (5 Joints)	0.67	0.87	1.19
Joint set K2 (15 Joints)	0.06	0.58	1.74

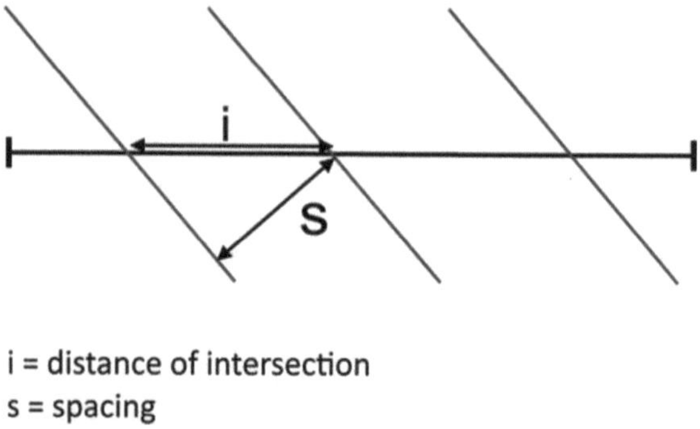

Fig. 6.8 Illustration of the intersection distance and the joint spacing related to the scanline (*black horizontal line*)

classified as "moderate" to "wide" according to ISRM (1978: 334). These values should be interpreted as indication values, since the joint spacing at the project site is partly irregular composed due to the variation in dolomite content.

The joint spacing represents the distance between two joints measured in a 2D-section in the recording plane of the scanline (not the distance of intersection). Due to the scanline providing a (in the current case) horizontal 2D-section through the rock mass, this joint spacing is not measured perpendicular to the joint surface, but perpendicular to the intersection line between the joint the scanline plane (Fig. 6.8). For very steep dipping joints this joint spacing corresponds to the true joint spacing in 3D-space.

The upper half trace length was recorded depending on the visible sections of the joint in upslope direction. As mentioned the terrain is partly vegetated and benched terrain. The lower half trace length was not recorded due to the recording-location situated on a forest road, below which no joint traces were visible. The additional data of joint roughness coefficient (JRC) and joint curvature were visually determined during the scanline analysis, incorporating an evident portion of subjectivity. Due to this fact the JRC for the stability analysis of the critical block was quantitatively and directly recorded at the failure surface. The curvature was categorized according to 5 classes where 1 is plane and 5 are most curved joints. This parameter is not supposed to have an increasing influence on the study results and thus was only recorded for completeness.

6.2.2 *Kinematic Analysis*

The data-basis for the kinematic analysis was provided by the discontinuity data of the "virtual" scanline. In total 105 planes were considered for the kinematic analysis taking the joint sets of Sect. 5.2.1 into account. The analyzed detachment modes include the planar failure, the wedge failure and the direct toppling mode, illustrated in the Schmidt Net, meaning the lower hemisphere equal area projection. For the kinematic analysis the following assumptions were taken in advance: The basic friction angle of 35° is based on literature values for limestone material (Cruden and Hu 1988; Heckmann et al. 2012; Hoek et al. 1998). It was not possible to perform rock shearing tests during the PhD period so the friction angle is based on the basic friction angle for this analysis. The assumed slope angle of 65° for the rock faces is based on information obtained from the slope inclination map generated in ArcGIS with 10° intervals.

The statistical data of the performed kinematic analysis are illustrated in Table 6.2. The potential for planar sliding is exclusively given for joints belonging to joint set K1, dipping parallel to the slope with a minor inclination angle. Considering a friction angle of 35°, 84 % of the joints belonging to set K1 are suggested to be hazardous for planar failure meaning 22 out of 26 joints. In relation to the total amount of recorded joints this would be about 22 %. This analysis result is in accordance with our observation from field work, since the failure surface of the critical block at the Wachterl-Horn and further blocks at the source area belong to joint set K1. The results of the limit equilibrium analysis linked to the kinematic analysis data are discussed in Sect. 6.1.4.

For wedge failure the critical intersections of joints dipping steeper than the critical friction angle (35°) and shallower than the slope angle are considered. The plot in Sect. 5.2.2 illustrates exceptionally the critical intersection points for wedge sliding being specified with 266 out of 1337 total joint intersections. Thus 19.9 % of the total amount of joint intersections is critical in terms of wedge failure. The intersection lines belong to the joint sets K1 and K2 providing potential rock wedges.

The evaluation of direct toppling contains the relevant intersections for freeing a certain block. In the current study this would be the intersections of joint set K2 and the bedding. Joint set K1 could provide an additional sliding plane. For direct toppling in total 186 intersections were considered, of which 167 (90 %) would be suggested to be critical.

The obtained results show that the dominant failure modes at the source area are suggested to be the direct toppling and the planar failure mode, the mode of wedge failure is suggested to be of minor relevance. The joint set K1 is of increased importance, since this orientation plays a major role for each of the considered detachment modes.

Table 6.2 Summary of the evaluation results of the kinematic analysis considering the detachment modes: planar failure, wedge failure and direct toppling

Failure type	Total planes	Total intersections	Critical amount (joints)	Critical amount (%)	Considered joints
Planar failure	26	–	22	85	Set K1
Wedge failure	–	1337	266	20	Bedding, K1, K2
Direct toppling	–	186	167	90	Bedding, K2

6.2.3 Recording of Block Axes at the Talus Slope

The sample area size of 20 by 20 m can be seen as an approximate value due to the partly difficult accessibility. We measured the sample area by extending two measuring tapes, one in upslope direction and one across the slope. Consequently variations in the sample area size of 1–3 m^2 should be taken into account. The sample areas were distributed over the slope almost equally to guarantee representative results for block dimensions in proximal, distal as well as lateral direction. In total five sample areas were evaluated accounting for the block dimensions (x-, y- and z-axis) as well as for the mean obstacle height (MOH). For the evaluation only blocks with a diameter larger than 0.1 m were taken into account due to two main reasons: blocks with a diameter smaller than 0.1 m will be pressed into the ground during the block-slope interaction and secondly the MOH of these blocks will have a minor effect in terms of slope roughness.

For the presentation of the block recording results it is to note that a factor of uncertainty should be taken into account for the interpretation of the results. The measuring uncertainty for the block diameters can be given with about ±0.02 m for the smaller block diameters. For the mean block diameters between 0.4 and 1.0 m an uncertainty of ± 0.05 m should be considered. Analogue to the mean diameter recording the uncertainty should be taken into account for the assessment of the mean obstacle height. For the MOH ranges of 0.01–0.1 m the uncertainty can be estimated with 0.005 m, for the MOH ranges between 0.1 and 0.4 m as 0.02 m and for the range between 0.4 and 1.0 m with ± 0.05 m.

For each sample area the total amount of recorded blocks, the amount of blocks with a mean diameter of 0.1 m and the amount of blocks showing a MOH are compared in Table 6.3. For sample area 1 the decrease between the total amount of recorded blocks and the blocks with a mean diameter larger than 0.1 m is evident. In this sample area the outcrop conditions at the talus material were very good, so the small block sizes have been recorded. At sample area 5 the outcrop conditions were bad due to an evident degree of vegetation and moss incrustation of the blocks. At this sample area the decrease between the total amounts of recorded

Table 6.3 Comparison of the recorded block amounts: the total number of recorded blocks, the amount of blocks with a mean diameter exceeding 0.1 m and the total amount of blocks providing a MOH in terms of slope roughness

Sample area	Total amount of recorded blocks	Total amount of blocks with a mean diameter larger than 0.1 m	Total amount of blocks with a recorded MOH
1	285	93	86
2	282	263	253
3	314	279	260
4	360	236	164
5	153	128	83

blocks is of minor importance whereas the decrease of blocks providing a mean obstacle height is recordable, which can also be recognized in sample area 4. In areas with an increasing vegetation cover and weathered rockfall material the recording can be complicated. Due to bad outcrop conditions the block dimensions can partly hardly be assessed and blocks with a small diameter can often rarely be recorded. Even larger blocks are vegetated to a large extent consequently providing a low MOH even in case of an increased mean diameter (Fig. 6.9).

Summing up, the results of Table 6.3 emphasize that not every recordable block at the talus slope provides a mean obstacle height in terms of slope roughness. Assessing the slope roughness via the mean block diameter instead of the mean obstacle height suggests overestimating the roughness parameters.

Comparing the results of the mean block diameters with the recorded MOHs, the most important fact to recognize is that the peak of the mean diameter distribution lies within a range of 0.1–0.4 m whereas the peak of the MOH distribution varies between 0.01 and 0.2 m. This indicates that the dominant part of blocks show a mean diameter which is about 44 % higher than the average value of the mean obstacle height.

In the following paragraph the results of the mean block diameter distributions will be discussed.

Fig. 6.9 The block on the photo shows a length of approximately 1.20 m. Due to weathering, surrounding block material and vegetation the MOH is only about 0.15 m

The sample areas 1, 4 and 5 provide similar results for the counting of the mean block diameter (Figs. 6.10, 6.11 and 6.12). The bar diagrams suggest a peak of the mean diameter distribution at a range between 0.1 and 0.4 m (about 85–88 % of the recorded blocks), where the dominant part of blocks provides a mean diameter between 0.1 and 0.3 m (about 65–67 % of the recorded blocks). The diameter classes of 0.41–0.5 m vary between 3.1 and 5.5 % and the class of 0.51–0.6 m ranges between 3.9 and 5.4 %. For the highest block diameter class the range between 0.6 and 1.3 m was grouped together due to the low amount of the blocks showing a dimension in this range. Sample area 1 shows a portion of 6.6 % and sample area 4 of 6.8 % of the recorded blocks with a mean diameter between 0.61 and 1.3 m. The percentage of the recorded blocks in a range between 0.61 and 1.3 m can be specified as 3.9 % for sample area 5.

The results of the sample areas 2 and 3 suggest similar results in terms of the mean diameter distribution (Figs. 6.13 and 6.14). Both distributions show an evident peak at a block diameter range between 0.21 and 0.4 m (about 55–61 % of the recorded blocks). A block amount of about 20 % (sample area 2) and about 18 % (sample area 3) shows a mean diameter range between 0.1 and 0.2 m. The third evident diameter class can be specified as the range between 0.31 and 0.4 m with a percentage of about 14 % in sample area 2 and 9 % in sample area 3. The block diameter ranges between 0.41 and 1.3 m are suggested to take a minor part of altogether 10 % (sample area 2) and 12.2 % (sample area 3).

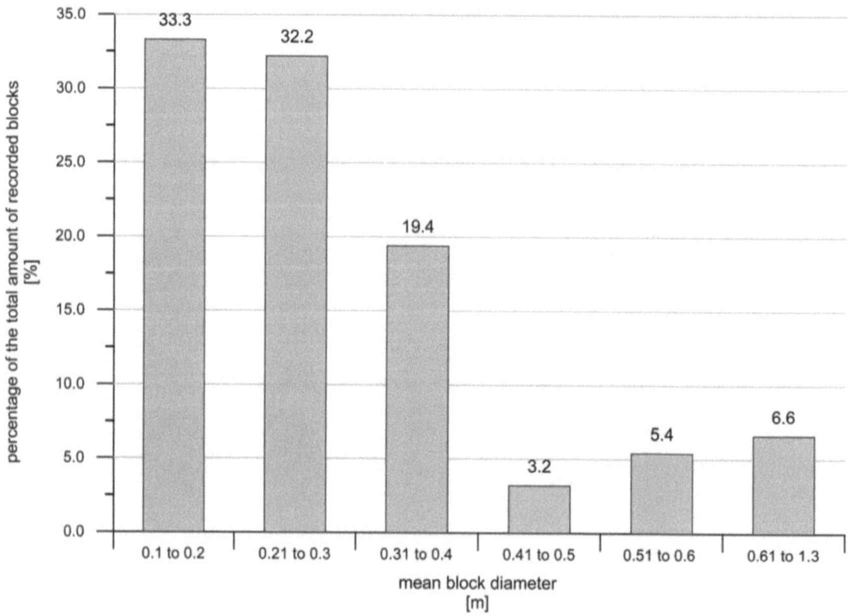

Fig. 6.10 The bar diagram shows the mean block diameters recorded at sample area 1 (*y-axis*), which is located most near to the valley bottom

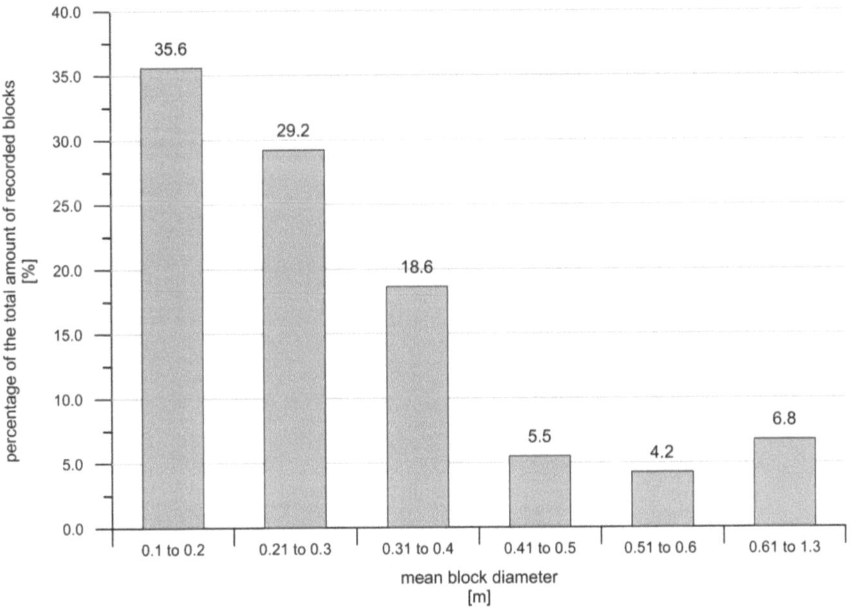

Fig. 6.11 The bar diagram shows the mean block diameters recorded at sample area 4 (*y-axis*), which is located at the middle part of the talus slope in the most south direction

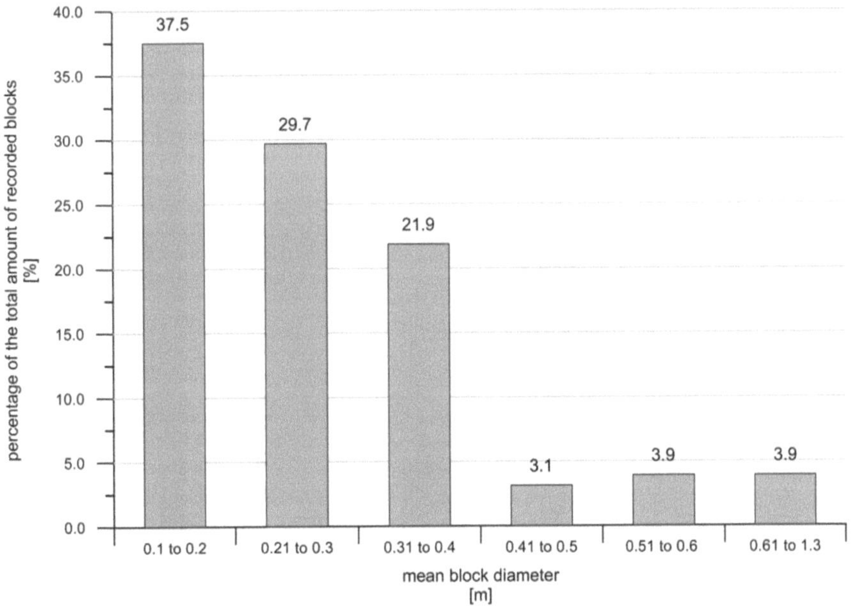

Fig. 6.12 The bar diagram shows the mean block diameters recorded at sample area 5 (*y-axis*), which is located most near to the source area

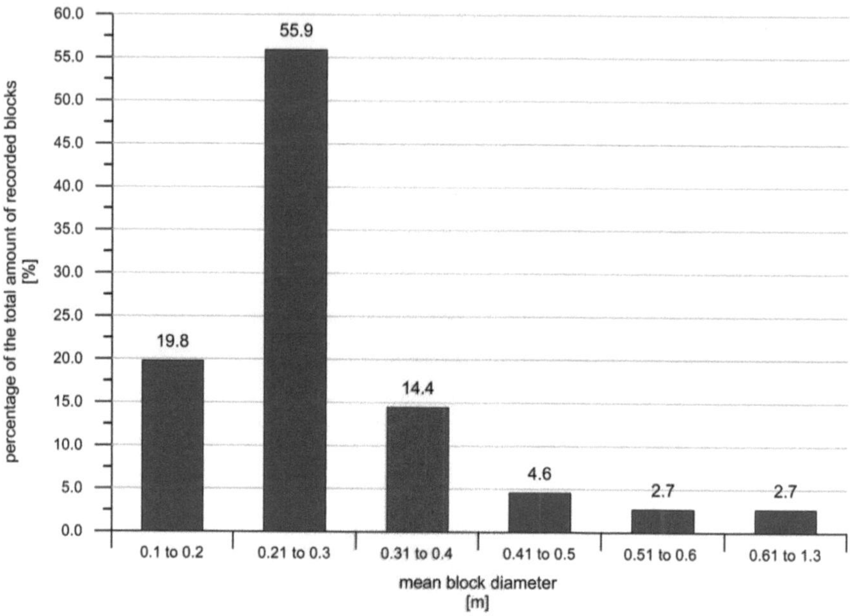

Fig. 6.13 The bar diagram shows the mean block diameters recorded at sample area 2 (*y-axis*), which is located at the middle part of the slope in the most northern direction

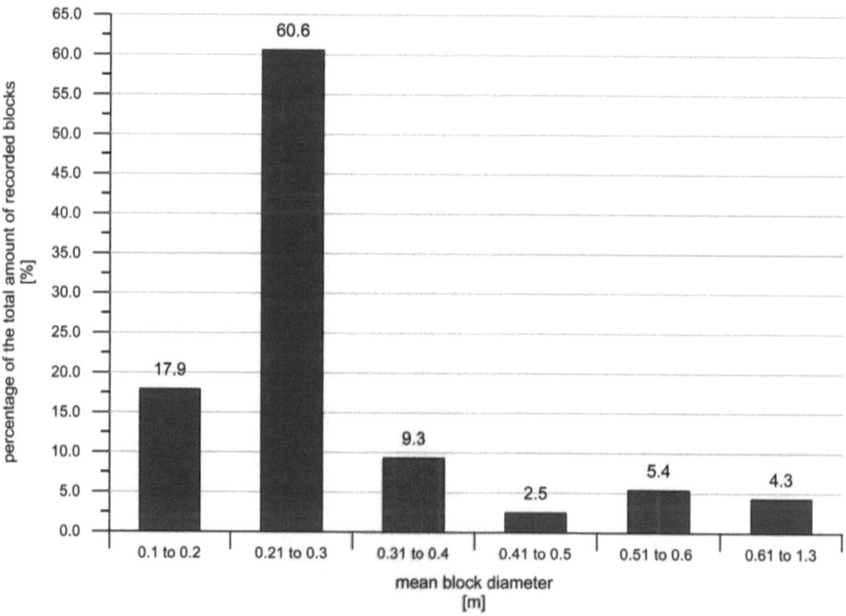

Fig. 6.14 The bar diagram shows the mean block diameters recorded at sample area 3 (*y-axis*), which is located at the middle part of the slope

In the following paragraph the mean obstacle height outcomes will be discussed.

Equal to the results of the mean diameter distributions, the sample areas 1, 4 and 5 are suggested to the grouped together in terms on the MOH-results (Figs. 6.15, 6.16 and 6.17). The sample areas 1 and 4 suggest a peak of the MOH distribution at a range between 0.01 and 0.2 m, with about 76–84 % of the total amount of considered blocks. The MOH range between 0.2 and 0.3 m can be specified as 11.6 % for sample area 1 and 4. For sample area 5 the dominant potion of blocks is represented by about 80 % of the blocks within a range of 0.01–0.1 m in terms of the MOH whereas the range between 0.1 and 0.2 m is represented by 10.8 %. For the sample areas 1 and 5 the MOH ranges between 0.3 and 1.0 m are suggested to represent a minor potion with 4.7 % for sample area 1, and 2.4 % for sample area 5. Only at sample area 4 this MOH range includes 12.2 % of the recorded blocks. The sample areas 2 and 3 again show similar outcomes for the distribution of the mean obstacle height, suggesting an exponential run of the envelope curve (Figs. 6.18 and 6.19). An evident maximum of the MOH distribution can be localized at a MOH range of 0.01–0.05 m with 65.6 % for sample area 2 and 57.3 % for sample area 3. With increasing MOH, the percentages of the block amounts decrease: with 21.3 % (sample area 2) and 20.8 % (sample area 3) for a MOH range between 0.051 and 0.1 m; 8.3 % (sample area 2) and 14.5 % for a MOH range between 0.1 and 0.2 m. The MOH range between 0.2 and 1.0 m can be specified with a total percentage of

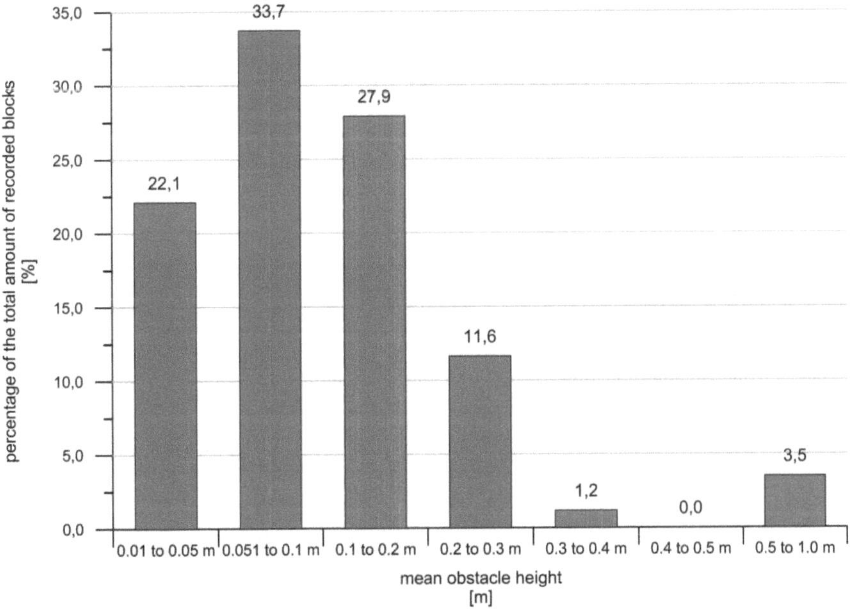

Fig. 6.15 The bar diagram shows the mean obstacle heights recorded at sample area 1, which is located most near to the valley bottom

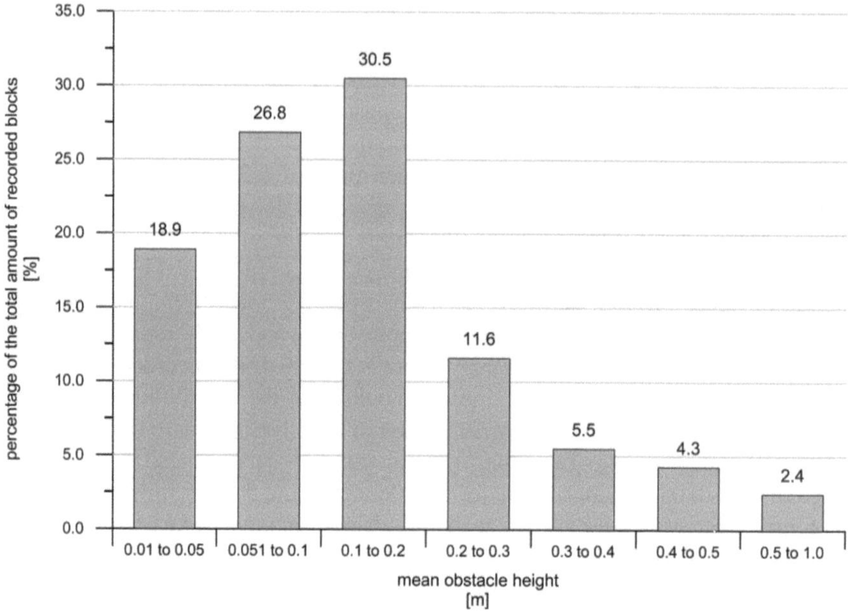

Fig. 6.16 The bar diagram shows the mean obstacle heights recorded at sample area 4, which is located at the middle part of the talus slope in the most south direction

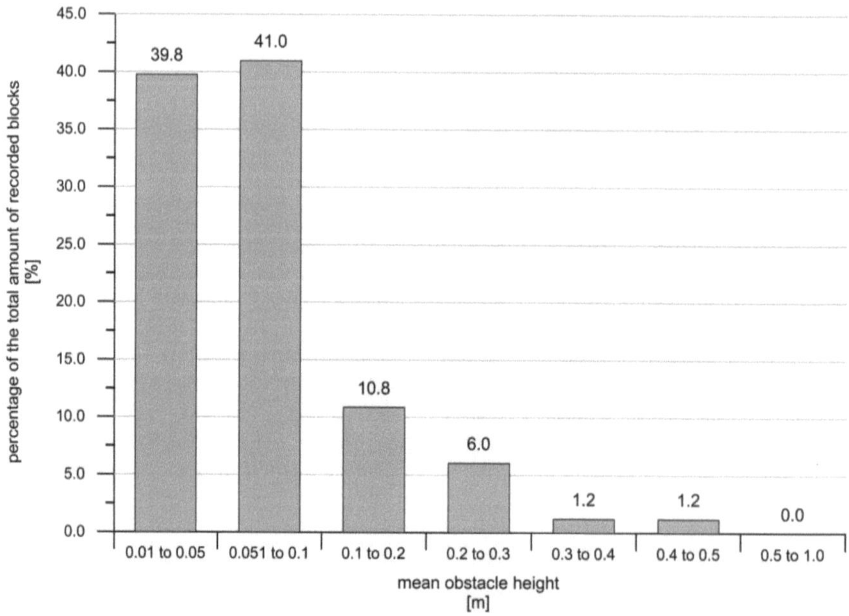

Fig. 6.17 The bar diagram shows the obstacle heights recorded at sample area 5, which is located most near to the source area

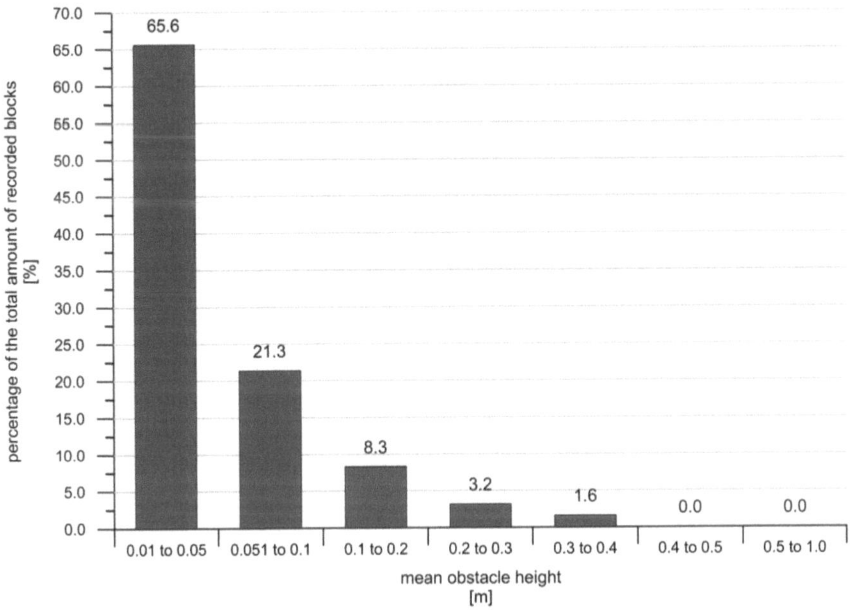

Fig. 6.18 The bar diagram shows the mean obstacle heights recorded at sample area 2, which is located at the middle part of the slope in the most northern direction

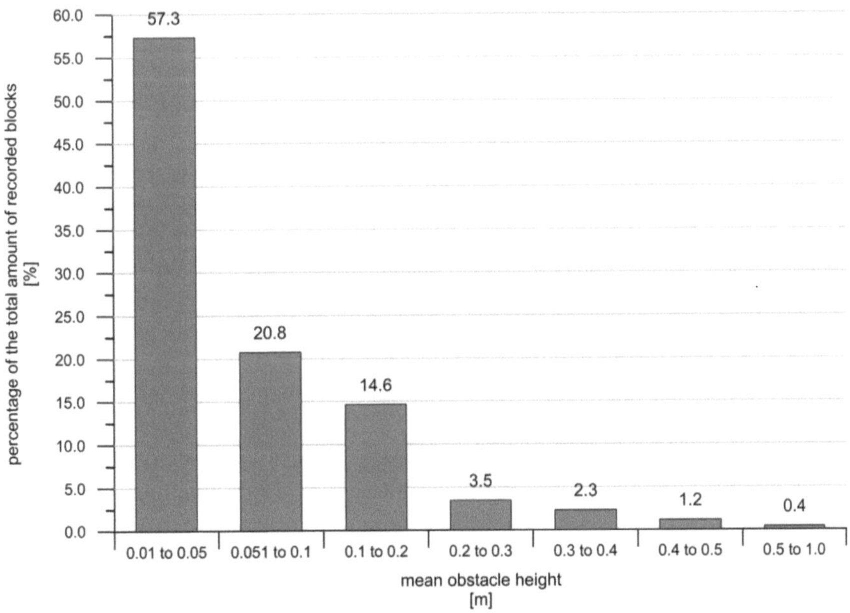

Fig. 6.19 The bar diagram shows the mean obstacle heights recorded at sample area 3, which is located at the middle part of the slope

4.6 % (sample area 2) and 7.4 % (sample area 3), which suggests to be the minor content.

Comparing the outcomes to the sample area locations the following summary can be given: The sample area 1 is located most near to the valley bottom, whereas the sample area 5 is situated near the source area. The sample areas 2, 3 and 4 cover the middle part of the talus slope. The mean diameter distributions of sample areas 1, 4 and 5 are suggested to be similar, even if the sample locations are not linked in terms of proximal or distal tendency at the talus slope. The sample areas 2 and 3 also provide similar diameter distributions. These two sample areas are situated next to each other at the talus slope. This is suggested to be one reason for the similarity. Nevertheless the outcomes suggest that there is no spatial tendency of sorting effect observable from the proximal to the distal edge of the talus slope.

At a non-forested slope a sorting effect could be assumed where the large blocks would descend farther at the talus slope than the smaller block volumes (Statham 1976). At the Weißwand we can observe two factors prohibiting this kind of sorting effect: the dense forest stand and the different altitude levels of rockfall source areas. Due to the different release altitudes, the blocks provide varying kinetic energies leading to different run-out distances. Additionally the forest stand leads to an evident loss of kinetic energy. Both influencing factors are suggested to result in a mixture of block volumes at the talus slope, where no sorting effect from the source area to the accumulation area at the talus slope is recognizable.

6.3 Rockfall Modelling

6.3.1 Rocky for 3D

The rockfall run-out for the entire project site was analyzed using the code Rockyfor3D. In the following paragraph the selected input parameters are discussed.

The assumed block volume-classes are based on quantitative data of rockfall recording (Sect. 5.2.3). The block class "Mean" (0.2 × 0.25 × 0.35 m) is represented by the dominant part of the recorded blocks at the talus slope approached as a frequent rockfall event at the Weißwand project site. The block classes "Min." (0.15 × 0.2 × 0.25 m) and "Max." (0.5 × 0.6 × 0.8 m) represent minor and major volumes based on the block-volume-evaluation, which can be followed in Sect. 6.2.3. The volume class "blocks" (0.8 × 1.0 × 1.2 m) corresponds to the rock volume approached by the State Environment Agency of Bavaria (Bayerisches Landesamt für Umwelt) by means of the hazard indication mapping of Bavaria. Referring to their data the Dachstein-Formation would belong to "volume class I", which would be considered as 1.2 × 1.2 × 1.2 m blocks (Bayerisches Landesamt für Umwelt 2014: 14). Since the rock dimensions were fully recorded at the project site the block axes of this class were adapted to the block shapes observed at the project site "Weißwand" as described above.

The slope roughness parameters (e.g. MOH) were recorded during field work and were afterwards validated by the parameter classes provided with the code Rockyfor3D (Dorren 2012). In this context it should be noted that the recorded mean obstacle height values presented in Sect. 6.2.3 are based on performing random sampling across the talus slope. The presented outcomes of the MOH-recording mirror the dimensions of single blocks at the sample area not representing an average mean obstacle height value of the entire homogenous areas related to the size of the homogenous area. Therefor the recorded MOH values have been validated and adapted using the recommendations of the Rockyfor3D manual (Dorren 2012).

Based on the data from field investigation one slope roughness dataset was created. A second parameter set was developed increasing the roughness for every homogenous area of 0.1 m, which had the following influence on the additional roughness (MOH)/block-diameter relation: For a mean diameter (y-axis) of 0.25 m = 40 %, for 0.35 m = 28.5 %, and for 0.6 m = 17 %. This relationship indicates an increasing effect on a smaller block diameter, which can be comprehended by the modelling results in Sect. 5.3.1.

Due to the densely forested project site it was decided to present the results based on the assumption of a forested slope. One example of comparing the rockfall run-out without forest stand to one with forest stand is illustrated for the volume class "Mean" in Sect. 5.3.1. The number of trees of the forest stand was recorded by counting tree crowns using aerophoto material (resolution of 20 cm per pixel) and validating this information in the field. The issue about this way of proceeding might be that only the highest trees visible on the aerophotos are considered. Small trees or brushwood are not incorporated in the forest input data. In case of additional brushwood or small tress the run-out might be additionally reduced. An actual approach of evaluating the forest stand from high resolution surface data is provided by ecorisQ, called FINT. This tool was provided one year after having performed the input data generation for the current model and is mentioned here for completeness.

In the following paragraph the modelling results are discussed.

Since the project site is densely forested the effect of the forest stand on the rockfall run-out (illustrated by the nr of deposited rocks) is visualized by the block class "Mean" in Sect. 5.3.1. The results suggest an decreasing run-out distance due to incorporating the forest stand, illustrated as the "nr. of deposited" rocks. The illustration of the "nr. of deposited" rocks is suggested to be a helpful tool for run-out mapping in terms of rockfall hazard estimation (Dorren 2012). The evaluation of this parameter in the different volume scenarios indicates on the one hand the increasing run-out distance with increasing rock-volume. On the other hand for each scenario, the results of the two slope roughness parameter sets are compared (Sect. 5.3.1). Due to the scale-relation between block diameter and additional roughness-factor, the effect of an increasing slope roughness increases with decreasing block diameter. An important fact to notice for the interpretation of the results of the "nr. of deposited" blocks is the boundary effect of the model. Every time a block in the simulation gets in contact with the model boundary, the block will immediately be stopped in this raster cell. In the current model, the north model boundary is further situated in an avalanche trench causing additional accumulation potential due to the morphological depression. Due to this fact the results of the number of deposited blocks suggest partly extraordinary high values for deposited blocks per square meter, for example 300–400 blocks for the maximum block class. We analyzed the parameter sets in ArcGIS and clustered the raster cells with deposit values between 0 and 200 blocks and 200–400 blocks per raster cell (Fig. 6.20). The outcomes indicate a minor variation between the model boundary effect (where blocks are immediately stopped when reaching the model boundary) and the high accumulation rates due to the avalanche trenches. A suggested way of proceeding with these effects could be the clipping of the model results in a GIS system. If the modelled area is selected wider than the area of required model data, this effect can be eliminated by clipping the extent of the model outcomes.

A further interesting effect is the increasing amount of accumulated blocks between the two rockcliffs. This area was defined as source area due to existing rockcliffs also visible on the hillshade map, even if the slope angle is decreasing to 30°–40°. Due to the decreasing slope angle the results suggest a release and source

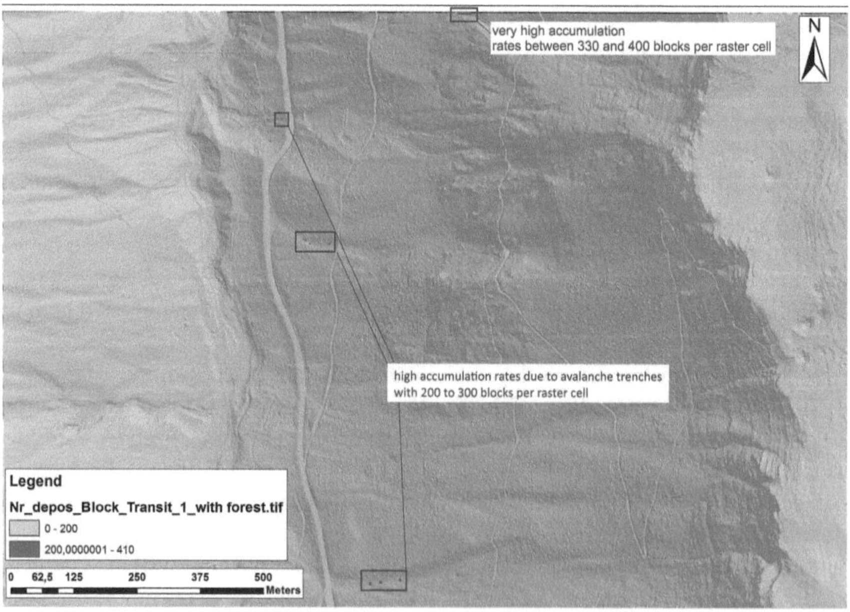

Fig. 6.20 For the discussion of the parameter set "Nr. of deposited" blocks we clustered the outcomes due to the accumulated block ranges 0–200 blocks and 200 to the maximum of 410 blocks. The outcomes show a variation between the high block accumulation rates at the avalanche trenches and the boundary effect of the model (where the blocks are stopped when reaching the model boundary)

area for the same part of the slope. This area was interpreted as zone of rolling and bouncing processes, where no free fall occurs due to the decreasing slope angle. This interpretation is further based on the shape of this accumulation area in southern direction. The edge of this accumulation zone strictly follows the shapefile of the defined source area (Fig. 5.15). A realistic approach might be to interpret this area as a primary but mainly as a secondary release area, where blocks stopped due to the decreasing inclination angle could be secondarily released.

The areas of windblow are a further factor influencing the slope roughness and thus the rockfall run-out. At the project site Weißwand the windblow areas are suggested to have an evident effect showing an increasing slope roughness due to root-plates and –stocks. In the current modelling the slope roughness was increased to absolute values of rg70 = 0.3 m, rg20 = 0.5 m and rg10 = 0.15 m. This assumption is based on field observation and the modification of the assessed obstacle heights to an average mean obstacle height (MOH) being relevant for a descending block (Sect. 2.2.3; Dorren 2012). Since up to now there are no published suggestions how to deal with windblow areas in 3D-modelling using the code Rockyfor3D, the applied values can be seen as a rough estimation, which could still be improved by quantitative recording during further work. Nevertheless the outcomes demonstrate that the remaining cut-off tree stems act similar to a rake retaining the block fragments in this area. Especially the large windblow area above the federal road extends over a large distance and thus shows a high retaining effect on the blocks. If the tree relicts would be completely removed, the rockfall run-out is suggested to be increasing due to missing forest stand and decreasing surface roughness. The outcomes of taking windblow areas into account in the code Rockyfor3D can be interesting for forest municipalities as well as for state agencies dealing with hazard assessment along infrastructure.

Summing up the outcomes of the Rockyfor3D model suggest a decrease of the rockfall run-out depending on the following parameters:

- With an incorporation the forest stand
- With decreasing the block volumes
- With increasing the slope roughness

The decreasing rockfall run-out implies also a decrease of the kinetic energies of the descending rocks. The increase of slope roughness can have different causes which are suggested be considered for the model set up: roughness caused by rock-fall material or increased roughness due to forest material, like fallen trees, root planes or root stocks.

In the following paragraph the outcomes of an assumed frequent event (block class “Mean”) will be compared to the outcomes of the maximum block size “blocks”.

The recorded block sizes of the talus material suggest a mean block diameter of 0.25 m in average. Due to the high percentage of this mean block diameter class in all 5 sample areas it was decided to assume this volume class (“Mean”) as an average frequent event at the Weißwand. The assumed block volumes of the Bavarian State Agency for Environment (Bayerisches Landesamt für Umwelt) suggest block volumes of 1.2 × 1.2 × 1.2 m as regular events for carbonate rocks of the Dachstein-Formation at the Weißwand (Bayerisches Landesamt für Umwelt 2014: 14). Due to the results of the quantitative block recording a mean block diameter of 1.2 m would correspond to 2.7–6.8 % of the total amount of recorded blocks at the sample areas across the talus slope. To compare the block classes of the Bavarian State Agency for Environment with the assumed block class "Mean", the rock volume class "Blocks" was taken into account (Sect. 4.3.2). The model outcomes suggest an evident variation in terms of block run-out as well as in terms of the kinetic energies. In case of the volume class "Mean" a minor amount of blocks will reach the federal road, if forest stand is taken into account. In case of the volume class "blocks" the federal road across the entire project site will be affected by descending blocks (under consideration of the forest stand). Considering the mean kinetic energies along the federal road, the outcomes of volume class "blocks" suggest average kinetic energy values about 1,600–1,800 kJ (requested from ArcGIS), whereas the kinetic energies of the block class "mean" can be described by 13–20 kJ (requested from ArcGIS). It is to note, that these results are to be considered as an approach for hazard assessment suggesting a guide value for the kinetic energies depending on the rock volume. Nevertheless the outcomes indicate a decrease of the kinetic energy along the federal road of about 99 % due to a variance in rock volume from 0.96 m^3 ("Blocks") to 0.0175 m^3 ("Mean"). Assuming an average load capacity of rockfall fences of about 2500 kJ the resulting kinetic energies of the volume class "blocks" could be absorbed by common rockfall fences.

In the context of hazard assessment the discussed results should emphasize the effect of quantitative determined rockfall parameters. The effect of varying rock volume shows not only an influence on the rockfall run-out but also on the modelling outcomes as basis parameters for mitigation measure design.

6.3.2 RAMMS::Rockfall

The aim of modelling with the code RAMMS::Rockfall during the project was to perform parameter studies in terms of fragmentation for the critical block at the Wachterl-Horn.

The results of the current parameter studies show an evident dependence of the following factors in terms of hazard assessment:

- The block volume (degree of fragmentation)
- The forest stand density

For incorporating the degree of fragmentation we took the assumption that the critical block would fragment with the first block-ground contact. Due to the opening width of about 1.0 m perpendicular to the failure surface this approach is suggested to be reasonable. Due to the current rockfall codes not being able to incorporate a "real" fragmentation during the descending process the mentioned assumption is crucial. With decreasing block volume the trajectory density of blocks reaching the federal road decreases, but still an evident amount of blocks reaches the road, for a non-forested slope. Even for the smallest editable rock volume in RAMMS::Rockfall, considered in the scenario "minimum rockfall" the blocks still affect the federal road in case of a non-forested slope. An interesting result for this scenario is that the trajectories come-up at the counter hillside. This effect only occurs for small rock volumes. A potential reason for this might be the small rocks tending to roll farther over the slope surface and coming-up at the counter hillside. With decreasing block volume it is suggested that the code provides the more reasonable results for forested steep alpine slopes in terms of run-out distance.

In the following paragraph the three main parameters (velocity, kinetic rock energy and jumping height) are discussed in dependence of the rock volume and the forest stand. The figures providing the modelling results in a 3D-view illustrated on a hillshade map are included in Sect. 5.3.2.

For the volume class "minimum fragmentation" the very high rock volume shows a dominant effect. Due to the rocks with a volume of 16–33 m^3 the forest stand has a minor influence since most of the tree stems are supposed to be felled. In the code RAMMS::Rockfall the forest stand is implemented via a drag force, depending on the effective forest stand height and the drag coefficient. For this volume scenario the resisting force of the forest drag seems to have a minor influence compared to the speedup-forces. Nevertheless the contact with the trees should have an effect on the energy loss and thus on the run-out distance. The results of the kinetic rock energies show a decrease in the kinetic energy of about 7.5 % for a forested slope compared to a non-forested one. The maximal jump heights of the descending rocks are very high with 32 m for a non-forested and even 40 m for a forested slope. The jump height in RAMMS::Rockfall is measured in vertical direction to the slope surface, which results in evidently increasing jump heights below steep rockcliffs, like in the current study (Bartelt et al. 2013: 11).The higher jump heights for the forested slope are suggested to results from the defined forest stand. The drag force provided by the forest stand was defined with a drag height of 5 m, which is far below the maximum and even the mean jump heights for this scenario. The increase of the maximum jumping heights of about 20 % is suggested to result from different trajectories at the rock cliff in the middle part of the slope. If a rock descends on a very steep passage this might result in an increasing vertical distance to the slope surface below the cliff and thus in an increased jump height. The jump heights (5–15 m) along the federal road would still exceed the height of most mitigation measures. The block dispersion along the federal road is one of the most important factors in terms of hazard assessment. For this volume scenario the run-out dispersion along the federal road can be specified with about 600 m measured in ArcGIS directly at the federal road.

For the volume class "mean fragmentation" the block dimensions were halved so the volumes range between 1.6 and 4.2 m^3. With the decreasing rock volume, the forest stand is suggested to have an increasing influence. The velocities for this block class are suggested to be reduced about 50 % (from 62 to 32 m/s). Due to the decrease in the block mass (and the force of gravity) the forest drag has a higher impact. In consequence of the velocity reduction the kinetic energy results suggest also a dependence on the forest stand. The maximum kinetic energy decreases about 65 % from a non-forested slope to a forested slope, which would be suggested to be an evident impact. The maximum jump height varies from 46 m for a non-forested slope to 11 m for the forested slope, which could be specified with about 75 %. A reason for this could be that in case of a forested slope the rocks are suggested to have a higher portion of ground contact compared to a certain descending distance. Even below the rock cliffs, the maximum jump height does not exceed 11 m, measured vertically to the slope surface. Along the federal road, the jump height is suggested to decrease to 2–3 m, which would be controlled installing rockfall fences. The block dispersion along the federal road is decreasing with minor block volume and the effect of the forest stand. For a non-forested slope the dispersion could be specified with approximately 680 m, whereas the dispersion in case of a forested slope decreases to about 390 m, measured in ArcGIS.

The volume class "maximum fragmentation" accounts for blocks of the quartered dimensions of the scenario "minimum fragmentation". The rock volumes vary between 0.2 and 0.7 m^3. In this volume class the effect of forest stand is even more increasing. The velocities of the non-forested slope decrease about 84 % to the forested slope (from 87 to 11 m/s). In consequence of the decreasing velocity, the kinetic energies are also reduced evidently from 3831 to 102 kJ, which could be specified as 97 %. This is suggested to be a very high proportion of energy reduction due to forest influence. Either the forest drag force is very high in comparison to the driving forces of the descending rock or the drag forces of the slope parameters (without forest stand) are low compared to the kinetic energies provided by this volume class. The maximum jumping height decreases about 94 % from the non-forested to the forested slope (from 50 to 2 m). As in the scenarios described above, the maximum jump heights occur below the steep rock cliffs at the middle part of the slope. At the federal road the jumping heights tend to decrease to about 1.0 m, which controlled by rockfall mitigation measures. The rocks considered during modelling the forested scenario are suggested to have more ground contacts than in the non-forested scenario. The rock-fall trajectories suggest that on a forested slope considering this volume class, the amount of rocks reaching the federal road is evidently decreasing and the rocks would not exceed the road. The trajectory dispersion measured along the federal road would be about 660 m in case of the non-forested slope and 190 m for the case of a forested slope, which corresponds to a decrease in block dispersion of 71 %.

The results of the volume class "minimum rockfall" contain the minimum spectrum of possible block sizes editable in RAMMS::Rockfall (0.1–1.0 m^3). Due to the very small block size the effect of the forest stand is visible most in the results. The velocity trajectories suggest a reduction of the maximum rock velocity up to 95 % (from 73 to 4 m/s). In consequence the kinetic energy is decreased drastically about 99.6 % (from 1230 to 4 kJ). This is suggested to be a very high proportion of energy reduction due to forest influence. Either the forest drag force is very high in comparison to the driving forces of the descending rock or the drag forces of the slope parameters (without forest stand) are low compared to the kinetic energies provided by the blocks of this volume class. The jump height is reduced about 98.7 % in case of a forested slope compared to a non-forested one (from 61 to 0.8 m). The jumping heights suggested for a non-forested slope are with 61 m very high for rocks providing a volume between 0.1 and 1 m^3. The dominant effect of this modelling scenario is the tremendous reduction of the run-out distance based on the forest stand data. The results suggest that in case of the non-forested slope the rocks would easily cross the road section whereas in case of a forested slope the rocks would tend to stop at the very upper part of the valley side. The dispersion range of the rocks across the federal road is suggested to be about 540 m in case of the non-forested slope. Based on the forest stand the rocks would tend to stop approximately 500 m above the federal road (measured from GIS data).

Summing up, the approached modelling scenarios suggest the following outcomes depending on the volume class as well as the forest stand:

- In case of a non-forested slope the blocks of all volume classes (even the smallest editable rocks!) would reach the federal road
- In case of a forested slope, the run-out distance as well as the portion of blocks reaching the road is evidently decreasing with reducing the rock volume
- Therefor the forest stand is suggested to have a high influence on the modelling results

With minor rock volume the discrepancy in terms of run-out between non-forested and forested slopes increases. For the smallest modelled rocks the discrepancy can be specified as about 99 % in terms of the kinetic energy, which is an enormous percentage. Due to RAMMS::Rockfall being a Beta version at the time this PhD thesis is composed, further development concerning the influence of slope parameters to the run-out distance could be expected.

Nevertheless the modelling outcomes are suggested to contribute to hazard assessment along vulnerable infrastructure in a certain manner. The design of mitigation measures depends on the following factors: the run-out distance (location), the kinetic energies (load capacity) and the jump height (height of the mitigation measure). This information can be obtained from the modelling results also for mid-magnitude events and different stages of potential fragmentation. For the current case we suggest that assuming a maximal fence-load capacity of 3000 kJ on a forested slope even the blocks of the volume class “mean fragmentation” (1.5–4.2 m^3) could be controlled by rockfall fences along the federal road due to average kinetic energies of 2350 kJ and jump heights of about 1.8–3 m. The scenario “mean fragmentation” suggests that the rock dimensions mapped directly at the critical block are halved.

An additional factor interesting for the application in terms of hazard assessment would be dispersion range of potential block fragments along a certain road section. Especially for the forested scenarios, the effect of the decreasing rock volume on the dispersion range is clearly visible; as for the scenario "mean fragmentation" the block dispersion is reduced by 35 % in comparison to the scenario "minimum fragmentation".

6.3.3 Summary-Discussion of 3D Rockfall Modelling

The following chapter should provide a frame discussion of the used rockfall codes in the context of application ranges.

The code Rockfor3D was used to analyze the rockfall hazard of the federal road across the entire project site. The code RAMMS::Rockfall however was selected to analyze the hazard potential of a mid-magnitude rockfall event based on the described case study. The code Rockyfor3D provides the possibility to perform regional probabilistic rockfall analysis based on parameter-maps (polygons) generated in a GIS system. The model approach is based on the coefficients of restitution requiring damping and roughness quantification as input-parameters characterizing the slope conditions. The characterization of the rock material is given by the block dimensions, the density and the block shape. All required basic parameters can be determined by field investigation.

The code RAMMS::Rockfall in contrast is based on a non-smooth contact dynamics approach referring to the law of Coulomb-friction, taking the block as a rigid body into account. The required model parameters include parameters the friction coefficient Mu, the coefficient of normal restitution as well as parameters concerning the reference system of the block (Eigenframe) and the slope reference system. As block parameters a defined block shape, the block volume and the material density would be required. For the user the block-parameters are quantifiable. However the slope parameters can be selected out of a drop down menu in the code according to qualitative categories ("extra soft" to "extra hard"), the quantification of the slope-parameter sets is demanding in terms of field investigation.

An important aspect in the context of hazard assessment is suggested to be the minimum block size editable in the applied code and the consideration of the block shape. In the code Rockyfor3D there is no minimum block size required. Referring to the recorded block data at the talus slope at the Weißwand we were able to model also the low magnitude-high frequency events (volume classes "Min." and "Mean"). In contrast in the code RAMMS::Rockfall a minimum rock volume of 0.1 m^3 is required, which corresponds approximately to a block with dimensions of $0.4 \times 0.4 \times 0.5$ m. Referring to the quantitatively determined block sizes of the current thesis, the minimum block size is suggested to be quite high. Nevertheless for the scope of modelling during this thesis this limitation of the block volume had no influence. The block sizes of the modelled case study exceeded the minimum block volume of 0.1 m^3. Nevertheless it is suggested to be desirable to provide the possibility of modelling smaller rock diameters using RAMMS::Rockfall, especially in terms of alpine hazard assessment.

In the context of block shape incorporation the following aspects are suggested to notice. In the code Rockyfor3D the rock is set to be considered as a sphere during the block-slope-interaction. The diameter of the sphere corresponds to the mean value of the larger two rock dimension specifications (d2 and d3). In contrast the rock shape is taken fully into account during the rock-ground-interaction in RAMMS::Rockfall. Actual research suggests that the block shape has a certain effect on the run-out distance (Glover et al. 2012).

Besides the rock and slope parameters the forest stand influences the rockfall run-out. The forest stand is specified in Rockyfor3D via a number of quantitative parameters like the number of trees per hectare, the diameter at breast height (DBH) and its standard deviation as well as the percentage of coniferous trees (Dorren 2012: 8). Due to the raster based approach these parameters are varied for each raster cell implying that a certain number of trees are placed within each pixel. By means of this method the forest stand is spatially modelled over the slope area near to a real forest stand. An important thing to notice is that by means of this approach a descending block must not necessarily hit a tree.

In RAMMS::Rockfall the retaining effect of the forest stand is implemented via an additional drag force affecting the descending block during each contact with the slope surface. The forest stand is quantified in the code with taking the tree height and the drag force into account, where the tree height is independent from the drag force. This approach implies (in contrast to Rockyfor3D) that every time a rock interacts with the slope surface and additional retaining drag force is applied, which causes a significant effect of the forest stand.

In the field of rockfall hazard assessment the effect of rock fragmentation is suggested to be crucial in terms of run-out prediction. At the current state of knowledge a potential degree of fragmentation cannot be modelled in terms of 3D run-out modelling. During the actual project a potential degree of fragmentation was considered for the run-out parameter studies of the mid-magnitude block at the Wachterl-Horn. This approach was based on the assumption that the rock would fragment during the first contact with the slope surface immediately after the detachment. Nevertheless there are approaches of analysing dynamic fragmentation during rockfall using discrete element approaches (Wang and Tonon 2011). This study emphasizes the importance of pre-existing fractures in terms of dynamic fragmentation during the rockfall process. The approach provides interesting information for well-known case studies like the critical block at the Wachterl-Horn. Due to dynamic fragmentation analysis not being in the focus of this research project this topic is beyond the scope of the current thesis. For the common practice of rockfall modelling the determination of pre-existing fractures in potential blocks is beyond the scope of knowledge, since the material conditions of the potential rocks are often not known.

References

Bartelt P, Buehler Y, Christen M, Dreier L, Gerber W, Glover J, Schneider M, Glocker CH, Leine R, Schreizer A (2013) RAMMS, User Manual v1.5 Rockfall.-83p. (pdf-version)
Barton NR (1973) A Review of a new shear strength criterion of rock joints. Eng Geol 7:287–332
Barton N, Choubey V (1977) The shear strength of rock joints in theory and practice. Rock Mech 10:1–54
Bayerisches Landesamt für Umwelt (LfU) [Hrsg.] (2014): Gefahrenhinweiskarte Alpen mit Alpenvorland—Landkreis Berchtesgadener Land.—78 S., München (Bayrisches L.-Amt für Umwelt, pdf-version)
Cruden DM, Hu XQ (1988) Basic friction angles of carbonate rocks from Kananaskis country, Canada. Bull Int Assoc Eng Geol 38:55–59
Dorren LKA (2012) Rocky for 3D (v5.0) revealed—Transparent description of the complete 3D rockfall model. ecorisQ paper (www.ecorisq.org) 31 p
Fecker E (1978) Geotechnical description and classification of joint surfaces. Bull Eng Geol 18:111–120
Glover J, Schweizer A, Christen M, Gerber W, Leine R, Bartelt P (2012) Numerical investigation of the influence of rock shape of rockfall run-out. In: EGU General Assembly 2012, held 22–27 April 2012 in Vienna, Austria p 11022
Goudie AS (2006) The Schmidt Hammer in geomorphological research. Prog Phys Geogr 30 (6):703–718
Heckmann T, Bimöse M, Krautblatter M, Haas F, Becht M, Morche D (2012) From geotechnical analysis to quantification and modelling using LiDAR data: a study on rockfall in the Reintal catchment, Bavarian Alps, Germany. Earth Surf Process Land 37:119–133
Hoek E, Marinos P, Benissi M (1998) Applicability of the geological strength index (GSI) for very weak and sheard rock masses. The case of the Athens Schist Formation. Bull Eng Geol Environ 57:151–160
ISRM—International Society for Rock Mechanics (1978) Suggested methods for the quantitative description of discontinuities in rock masses—commission on standardization of laboratory and field tests. Int J Rock Mech Min Sci Geomech Abstr 15(4):319–368
Krahn J, Morgenstern NR (1979) The ultimate reistance of rock discontinuities. Int J. Rock Mech Min Sci Geomech Abstr 16:127–133
McCarroll D (1997) A template for calculating rock surface roughness. Earth Surf Process Land 22:1229–1230
McCarroll D, Nesje A (1996) Rock surface roughness as an indicator of degree of rock surface weathering. Earth Surf Process Land 21:963–977
Priest SD (1993) Discontinuity analysis for rock engineering. Chapman & Hall, London, p 473
Priest SD, Hudson JA (1981) Estimation of discontinuity spacing and trace length using scanline surveys. Int J Rock Mech Min Sci Geomech Abstr 18:183–197
Statham I (1976) A scree slope rockfall modell. Earth Surf Process 1:43–62
Wang Y, Tonon F (2011) Discrete element modeling of rock fragmentation upon impact in rockfall analysis. Rock Mech Rock Eng 44(1):23–35

Chapter 7
Conclusion

The current PhD thesis is divided into three main chapters being framed by the topic of rockfall hazard assessment. At Sect. 2.3, at the introduction part, key questions were provided for every section of the thesis. In the following chapter answers to the specified questions will be suggested analogue to the question-structure of Sect. 2.3. In the second section a summary conclusion will be given to put the examined topics in the context of rockfall hazard assessment.

1. Detachment mechanics at the source area:

    (a) For the case study of the critical block at the Wachterl-Horn a mechanical model based on detailed field investigation was suggested. The dimensions of the block and in consequence a potential rock volume was determined. Due to the unique occasion of an accessible failure surface underneath the block, the contact area between block and rock mass was recorded in detail. At the failure surface sheared rock material (rock flour) was observed providing certain evidence of an active rock-bridge cracking. A map of the failure cave was created illustrating the dimensions as well as the contact conditions between block and rock mass. In addition 4 cross sections through the failure cave were provided; visualizing that the shape of block and rock mass match to a certain extend. Based on field analysis a mechanical model for a potential failure scenario was developed. The model suggests that a failure scenario would depend on two key-blocks: The critical ("key") block and the toe-block. The toe-block is suggested to rotate over the bottom edge causing a lift-movement of the active block perpendicular to the failure surface. In addition the joint persistence was recorded from outside and inside to obtain knowledge about a potential degree of fragmentation in case of failure.

    (b) The shear parameters for this study were recorded directly at the failure surface underneath the critical block. The joint roughness coefficient was recorded along 4 cross sections distributed equally along the failure cave in a up to 5 mm level of detail, being corrected due to the length of the failure surface (Barton and Choubey 1977). The failure surface is not only characterized by roughness but also by an undulation over the failure surface. The joint wall compressive strength (JCS) was determined performing 200

B. Sellmeier, *Quantitative Parameterization and 3D-run-out Modelling of Rockfalls at Steep Limestone Cliffs in the Bavarian Alps*, Springer Theses,
DOI 10.1007/978-3-319-24510-2_7

clustered Schmidt-Hammer tests in 20 test series (10 tests per series). The results of the test series show mean values varying between 20 and 70 MPa for the partly wet and with sinter crusts covered failure surface. The field recording was validated by performing uniaxial compressive strength tests (UCS-tests) in the laboratory. The UCS values vary between 82 and 115 MPa and are thus about 54 % in average higher than the JCS values.

(c) The results from shear parameter recording were taken into account for a stability analysis for a block on an inclined plane. For the limit equilibrium analysis two scenarios were considered, due to the divergence of the morphologically visible block at the field site and the dimensions of the failure surface across the slope (which is extending farther). This fact leads to two different block volumes, which were both taken into account. The results from the limit equilibrium analysis illustrate that for an assumed basic friction angle of 35° and a mean slope angle of 55° (without taking any cohesion into account), the factor of safety is suggested to be 0.5, instable. Due to this result, an evident part of the resisting forces has to be provided by (1) the toe of the block, (2) potential rock-bridges and (3) additional roughness contacts. The further analysis focused on either providing the additional resisting forces via cohesion or friction. For the first we assumed a labile stage of stability for a planar case of failure (1.0) and back-calculated the required cohesion, which could be specified as 37 kN/m^2. For the latter we assumed the difference between the slope and the basic friction angle as necessary to be compensated by the dilation angle "i". The relation between the JRC and the dilation angle "i" was plotted for range values of the assumed JCS results. The outcomes suggest that a JRC of at least 8 would be required to compensate the driving forces under the given conditions.

2. Quantitative magnitude assessment and detachment processes

(a) For the entire project site an analysis of the discontinuity sets was performed. The main discontinuity sets (Bedding 080/40, K1 240/55 and K2 330/85) provide an orthogonal joint system at the source area. The rock face at the source area is in average oriented with 270/65. The main part of the joints was recorded along a "virtual" scanline following a forest road crossing the entire project site. Along this "virtual" scanline, in total 105 joints were recorded providing the data basis for the subsequent kinematic analysis. The kinematic analysis covers the following modes of failure: planar sliding, wedge failure and direct toppling. The results illustrate, that the joint set K1 (dipping parallel to the slope) is exposed to planar sliding, which is in accordance with our observation from field work (as the failure surface of the critical block belongs to this joint set). The discontinuities being exposed to wedge failure belong to the joints sets K1 and K2; with K1 dipping parallel to the slope and K2 steeply drawing through the rock mass. The mode of direct toppling is influenced by the joint set K2 and the bedding, dipping versus the slope. These two joint sets free the potential blocks from the environing rock mass, where the joint set K1 could function as an additional sliding plane for certain cases. At the critical block at the Wachterl-Horn two 20 m scanlines were recorded to perform detailed analysis of the discontinuity system at the vicinity of the block. The recorded parameters included the joint spacing, the upper half trace length, the joint roughness coefficient (JRC) and the curvature each recorded joint. The most interesting parameter during the current study was the evaluation of the joint spacing to amend the rock volume data at the source area.

(b) For determining potential rockfall volumes in a quantitative and structured way, we recorded the block axes and the mean obstacle height (MOH) at the talus slope. The sample areas measured 20 by 20 m and were distributed over the talus slope in proximal and distal as well as lateral direction to obtain statistical knowledge about the block dimension distribution. The results show that the dominant portion of rockfall material would be provided in a mean diameter range of 20–30 cm. The maximum recorded mean block diameters at the talus slope would vary between 60 and 130 cm, providing 5–8 % of the rockfall material. The counting results of the talus slope do not show evident variation between the sample areas.

(c) The results of the block recording were used as data basis for the block dimensions using the code Rockyfor3D. The rock volume classes namely Min. (0.15 × 0.2 × 0.25 m), Mean (0.2 × 0.25 × 0.35 m) Max. (0.5 × 0.6 × 0.8 m) are based on the recording of block dimensions, specifically on the mean diameter (y-axis). The dominant portion of recorded rocks was assumed to correspond to the volume class "Mean" implying a regular frequency event. The volume classes "Min." and "Max." were selected referring to the decreasing and increasing percentages of the block diameter referring to the evaluated data. The volume class "blocks" was created referring to suggested block volumes of the Bavarian Environment Agency (Bayerisches Landesamt für Umwelt 2014: 14). The block volumes of 1.2 × 1.2 × 1.2 m were approached for carbonate rocks of the Dachstein-Formation and applied to 3D-rockfall modelling for the hazard indication map of Bavaria (Germany). The block dimensions of the hazard indication map were adapted to the block shapes recorded during field investigation. The block shapes presented in the modelling matrix (Fig. 4.15) is a combination of block dimensions determined by field work and the suggested block dimensions of the Bavarian hazard maps. The results of the recorded mean obstacle height (MOH) have to be interpreted in relation to the recorded number of blocks and the slope roughness suggestions given by Dorren (2012). The considered sample areas of 20 × 20 m represent only a small excerpt of the talus material, whereas the MOH roughness values (rg 70, 20 and 10) are average values for each homogenous area at the talus slope. Thus the recorded mean obstacle height values provide a certain range for the MOH, which is suggested to be verified by the recommendation by the manual of Rockyfor3D (Dorren 2012).

3. Run-out modelling
   (a) The influence of slope roughness on the run-out distance was analyzed using the code Rockyfor3D for the entire project site. The basic data set of slope parameters was based on the parameters recorded by field investigation. For the enhanced roughness parameter set the roughness was increased adding 0.1 m to every roughness class (rg70, rg20 and rg10) and each homogenous area. The consequence is an increasing influence of roughness with decreasing rock dimensions. In the presented results the outcomes based on the determined field parameters (Transit_1) are compared to the parameter set based on an increased roughness (Transit_2). The results provide information about the number of deposited rocks (meaning the accumulated rock per raster cell), the mean kinetic energies as well as the maximum jump height. Especially in the upper part of the slope, between the two rock cliffs, the increasing roughness is suggested to have an evident effect on the rockfall run-out due to the decreasing slope inclination angle. In this area a dominant part of the released rocks is accumulated especially for small volume classes.
   (b) The effect of windblow areas was analyzed by modelling using Rockyfor3D. The results were presented for the volume class "blocks", since this scenario would correspond to a maximum mean frequent event. The extension of the windblow areas was determined by field investigation and aerophoto analysis. Two major windblow areas were located at the project site: one directly above the federal road extending across the slope and one at a middle altitude level of the talus slope extending at the south part of the talus slope. The effect of root stocks and root planes was considered adding an additional roughness to the mean obstacle heights of the talus slope (absolute values of rg70 = 0.3 m, rg20 = 0.5 m and rg10 = 0.15 m). This increased slope roughness results in a retaining effect on the descending blocks illustrated in Fig. 5.38. The windblow areas seem to act like a grate keeping the descending rocks back in this area, which results in a decreasing amount of blocks reaching the federal road.
   (c) For the critical block at the Wachterl-Horn we considered varying stages of potential fragmentation scenarios using the code RAMMS::Rockfall. The rock fragmentation cannot be considered in the used rockfall code. Due to this fact the assumption of a fragmentation during the first block—rock mass interaction was taken into account, which is suggested to be a reasonable approach for the current case due to the mechanical interpretation (Sect. 5.1.6). The rockfall scenarios were chosen according to four volume classes: the block dimensions recorded in the field, the halved block dimensions, the quartered block dimensions and the minimal in RAMMS:: Rockfall editable rock volumes (as a reference). Each volume scenario was

modelled with and without taking the forest stand into account. The outcomes show a dependence of the rockfall run-out on the volume class especially for the forested scenarios. The characteristic parameters (velocity, kinetic rock energy and jump height) indicate an evident dependence on the forest stand and the rock volume. For a mean degree of fragmentation (halved block dimensions 1.5–4.2 m^3), the kinetic rock energy is reduced about 66 % in case of the forested slope. For a high assumed degree of fragmentation (quartered block dimensions, 0.2–0.7 m^3) the kinetic rock energy decreases even about 97 %. As an interesting outcome for hazard assessment, the dispersion range of the descending blocks can be followed. With decreasing rock volume a decrease in length of the affected road section is recognizable. In case of a very high degree of fragmentation ("minimal rockfall") on the forested slope the blocks are suggested not to reach the federal road at all.

Summing up the current PhD thesis provides an integral approach of quantitative parameter determination in rockfall hazard assessment. The thesis covers the mechanical investigation of the detachment process of mid-magnitude events by means of a selected case study. Further input was given to the quantitative determination potential block volumes due to rock recording at the talus slope and scanline analysis at the source area. Based on the input parameters resulting from field work 3D rockfall modelling with the focus on hazard assessment was performed.

References

Barton N, Choubey V (1977) The shear strength of rock joints in theory and practice. Rock Mech 10:1–54

Bayerisches Landesamt für Umwelt (LfU) [Hrsg.] (2014) Gefahrenhinweiskarte Alpen mit Alpenvorland—Landkreis Berchtesgadener Land. 78 S., München (Bayrisches L.-Amt für Umwelt, pdf-version)

Dorren LKA (2012) Rocky for 3D (v5.0) revealed—Transparent description of the complete 3D rockfall model. ecorisQ paper (www.ecorisq.org) 31 p

Outreach

The current thesis provides the basis for further analysis in terms of field investigation as well as in terms of rockfall modelling. In the following chapter further ideas are described which could contribute to an integral way of rockfall hazard assessment.

The case study of the critical block suggests a way of field work for providing a reasonable data basis for analyzing the detachment process. This data could further be applied to a numerical detachment analysis using a distinct element code like UDEC or 3DEC. This modelling could provide further knowledge about the stage of stability regarding the assumption that the system could be simplified to a 2 block problem on an inclined plane. To obtain further knowledge about the stage of stability it would be interesting to ascertain the amount of rockbridges underneath the block. Therefore an interesting site investigation would be a ground penetrating radar system to quantify at which locations integral rock-rock-contacts are remaining. To confirm the outcomes of the stability analysis in terms of the friction angle it would be desirable to perform direct rock shear tests and validate the performed stability analysis.

To improve the assessment of rockfall magnitudes laserscanning and photogrammetry could be a powerful tool. Due to the high vegetation rate and the steep inclination angle of the whole slope at the valley side during this project, we were not able to perform laser scanning. At other accessible source areas the information of the discontinuity sets at the source areas could be directly compared to the information about the blocks recorded at the talus slope and thus provides an average value for a potential degree of material fragmentation. This way of evaluation could provide a reasonable data basis for potential rock volume evaluation.

The field of run-out modelling still offers potential for improving integer rockfall hazard assessment. The results of the modelling using Rockyfor3D provide the possibility of performing a pixel-based data request in ArcGIS for the design of mitigation measures. Since the output data are imported as raster data, every pixel contains quantitative information. Along infrastructure it could be interesting to request the mean jumping heights in terms of rockfall fence design to obtain statistical data at which part of the road section the maximum jump heights occur. The same procedure could be provided for the data set of "nr. of deposited" rocks and

B. Sellmeier, *Quantitative Parameterization and 3D-run-out Modelling of Rockfalls at Steep Limestone Cliffs in the Bavarian Alps*, Springer Theses,
DOI 10.1007/978-3-319-24510-2

the mean kinetic energies to analyze where and to which extend the most rockfall trajectories would affect the infrastructure.

The effect of windfall areas on the rockfall run-out could be optimized by analyzing the airborne laserscanning data by subtracting the first pulse and last pulse data, where the first pulse data would correspond to the Digital Terrain Model (DTM, including the vegetation cover) and the last pulse data would correspond to the Digital Elevation Model (DEM, excluding the vegetation cover).

The rockfall trajectory data of the code RAMMS::Rockfall are suggested to contribute in an enhanced way to the analysis of mid-magnitude events. Due to the possibility of directly editable block volumes and shapes, detailed case studies are possible. An important question in terms of hazard assessment along infrastructure is the extent of the rock dispersion along the affected section of infrastructure. The trajectories could be integrated into a GIS system and the information about run-out and dispersion range can be directly determined. For future project sites, where high frequency/low magnitude rockfalls and Mid-Magnitude events occur, a combined hazard assessment using Rockyfor3D and RAMMS::Rockfall could be interesting. The results of both codes could be integrated into a Geodatabase system using GIS analysis tools for run-out and mitigation measure design.

MIX
Papier aus verantwortungsvollen Quellen
Paper from responsible sources
FSC® C105338

If you have any concerns about our products,
you can contact us on
ProductSafety@springernature.com

In case Publisher is established outside the EU,
the EU authorized representative is:
Springer Nature Customer Service Center GmbH
Europaplatz 3, 69115 Heidelberg, Germany

Printed by Libri Plureos GmbH
in Hamburg, Germany